D0672230

STATISTICAL MODELING and ANALYSIS

for DATABASE MARKETING

Effective Techniques for Mining Big Data

STATISTICAL MODELING and ANALYSIS

for DATABASE MARKETING

Effective Techniques for Mining Big Data

Bruce Ratner

CHAPMAN & HALL/CRC

A CRC Press Company

Boca Raton   London   New York   Washington, D.C.

Library of Congress Cataloging-in-Publication Data

Ratner, Bruce.
    Statistical modeling and analysis for database marketing : effective techniques for mining
big data / Bruce Ratner.
    p. cm.
    Includes bibliographical references and index.
    ISBN 1-57444-344-5 (alk. paper)
    1. Database marketing--Statistical methods. I. Title

HF5415.126.R38 2003
658.8'--dc21   2002041714

Visit the CRC Press Web site at www.crcpress.com

Dedication

This book is dedicated to:

My father Isaac — my role model who taught me by doing, not saying.

My mother Leah — my friend who taught me to love love and hate hate.

My daughter Amanda — my crowning and most significant result.

Preface

This book is a compilation of essays that offer succinct, specific methods for solving the most commonly experienced problems in database marketing. The common theme among these essays is to address each methodology and assign its application to a specific type of problem. To better ground the reader, I spend considerable time discussing the basic methodologies of database analysis and modeling. While this type of overview has been attempted before, my approach offers a truly nitty-gritty, step-by-step approach that both tyros and experts in the field can enjoy playing with. The job of the data analyst is overwhelmingly to predict and explain the result of the target variable, such as RESPONSE or PROFIT. Within that task, the target variable is either a binary variable (RESPONSE is one such example) or a continuous variable (of which PROFIT is a good example). The scope of this book is purposely limited to dependency models, for which the target variable is often referred to as the "left-hand" side of an equation, and the variables that predict and/or explain the target variable is the "right-hand" side. This is in contrast to interdependency models that have no left- or right-hand side, and are not covered in this book. Since interdependency models comprise a minimal proportion of the analyst's work load, the author humbly suggests that the focus of this book will prove utilitarian.

Therefore, these essays have been organized in the following fashion. To provide a springboard into more esoteric methodologies, Chapter 2 covers the correlation coefficient. While reviewing the correlation coefficient, I bring to light several issues that many are unfamiliar with, as well as introducing two useful methods for variable assessment. In Chapter 3, I deal with logistic regression, a classification technique familiar to everyone, yet in this book, one that serves as the underlying rationale for a case study in building a response model for an investment product. In doing so, I introduce a variety of new data mining techniques. The continuous side of this target variable is covered in Chapter 4. Chapters 5 and 6 focus on the regression coefficient, and offer several common misinterpretations of the concept that point to the weaknesses in the method. Thus, in Chapter 7, I offer an alternative measure — the predictive contribution coefficient — which offers greater utility than the standardized coefficient.

Up to this juncture, I have dealt solely with the variables in a model. Beginning with Chapter 8, I demonstrate how to increase a model's predictive power beyond that provided by its variable components. This is accomplished by creating an interaction variable, which is the product of two or more component variables. To test the significance of my interaction variable, I make what I feel to be a compelling case for a rather unconventional use of CHAID. Creative use of well-known techniques is further carried out in Chapter 9, where I solve the problem of market segment classification modeling using not only logistic regression but CHAID as well. In Chapter 10, CHAID is yet again utilized in a somewhat unconventional manner — as a method for filling in missing values in one's data. In order to bring an interesting real-life problem into the picture, I wrote Chapter 11 to describe profiling techniques for the database marketer who wants a method for identifying his or her best customers. The benefits of the predictive profiling approach is demonstrated and expanded to a discussion of look-alike profiling.

I take a detour in Chapter 12 to discuss how database marketers assess the accuracy of the models. Three concepts of model assessment are discussed — the traditional decile analysis, as well as two additional concepts I introduce: precision and separability. Continuing in this mode, Chapter 13 points to the weaknesses in the way decile analysis is used, and instead offers a new approach known as the bootstrap for measuring the efficiency of database models. Chapter 14 offers a pair of graphics or visual displays that have value beyond the commonly used exploratory phase of analysis. In this chapter, I demonstrate the hitherto untapped potential for visual displays to describe the functionality of the final model once it has been implemented for prediction.

With the discussions described above behind us, we are ready to venture to new ground. In Chapter 1, I talked about statistics and machine learning, and I defined that statistical learning is the ability to solve statistical problems using nonstatistical machine learning. GenIQ is now presented in Chapter 15 as such a nonstatistical machine learning model. Moreover, in Chapter 16, GenIQ serves as an effective method for finding the best possible subset of variables for a model. Since GenIQ has no coefficients — and coefficients are the paradigm for prediction — Chapter 17 presents a method for calculating a quasi-regression coefficient, thereby providing a reliable, assumption-free alternative to the regression coefficient. Such an alternative provides a frame of reference for evaluating and using coefficient-free models, thus allowing the data analyst a comfort level for exploring new ideas, such as GenIQ.

About the Author

Bruce Ratner, Ph.D., is President and Founder of DM STAT-1 CONSULT-ING, the leading firm for analysis and modeling in the database marketing industry, specializing in statistical methods and knowledge discovery and data mining tools. Since 1986, Bruce has applied his expertise in the areas of marketing research, banking, insurance, finance, retail, telecommunications, mass and direct advertising, business-to-business, catalog marketing, E-commerce and Web-mining.

Bruce is active in the database marketing community as the instructor of the advanced statistics course sponsored by the Direct Marketing Association, and as a frequent speaker at industry conferences. Bruce is the author of the *DM STAT-1 Newsletter* on the Web, and many articles on modeling techniques and software tools. He is a co-author of the popular text book *The New Direct Marketing*, and is on the editorial board of *The Journal of Database Marketing*.

Bruce holds a doctorate in mathematics and statistics, with a concentration in multivariate statistics and response model simulation. His research interests include developing hybrid modeling techniques, which combine traditional statistics and machine learning methods. He holds a patent for a unique application in solving the two-group classification problem with genetic programming.

Acknowledgment

This book like all books — except the Bible — was written with the assistance of others. First and foremost, I acknowledge HASHEM who has kept me alive, sustained me, and brought me to this season.

I am significantly grateful to my friend and personal editor Lynda Spiegel, who untwisted my sentences into free-flowing statistical prose. I am appreciative of the hard work of Paul Moskowitz, who put the tables and figures into a useable format.

I am indebted to the staff of CRC Press for their excellent work (in alphabetical order): Gerry Jaffe, Project Editor; Allyson Kline, Proofreader; Suzanne Lassandro, Copy Editor; Shayna Murry, Designer; Rich O'Hanley, Publisher; Will Palmer, Prepress Technician; Pat Roberson, Project Coordinator; and Carol Shields, Typesetter.

Last and assuredly not least, it is with pride that I acknowledge my daughter Amanda's finishing touch by providing the artwork image for the book cover.

Contents

1

Introduction

Whatever you are able to do with your might, do it.

— **Koheles 9:10**

1.1   The Personal Computer and Statistics

The personal computer (PC) has changed everything — both for better and for worse — in the world of statistics. It can effortlessly produce precise calculations, eliminating the computational burden associated with statistics; one need only provide the right questions. With the minimal knowledge required to program it, which entails telling it where the input data reside, which statistical procedures and calculations are desired, and where the output should go, tasks such as testing and analysis, the tabulation of raw data into summary measures, as well as many others are fairly rote. The PC has advanced statistical thinking in the decision making process, as evidenced by visual displays, such as bar charts and line graphs, animated three-dimensional rotating plots, and interactive marketing models found in management presentations. The PC also facilitates support documentation, which includes the calculations for measures such as the mean profitability across market segments from a marketing database; statistical output is copied from the statistical software, then pasted into the presentation application. Interpreting the output and drawing conclusions still require human intervention.

Unfortunately, the confluence of the PC and the world of statistics have turned generalists with minimal statistical backgrounds into quasi-statisticians, and affords them a false sense of confidence because they can now produce statistical output. For instance, calculating the mean profitability is standard fare in business. However, mean profitability is not a valid summary measure if the individual profit values are not bell-shaped; this is not uncommon in marketing databases. The quasi-statistician would doubtless

1

not know to check this assumption, thus rendering the interpretation of the mean value questionable.

Another example of how the PC fosters a quick and dirty approach to statistical analysis can be found in the ubiquitous correlation coefficient, which is the measure of association between two variables and is second in popularity to the mean as a summary measure. There is an assumption (which is the underlying straight-line relationship between the two variables) that must be met for the proper interpretation of the correlation coefficient. Rare is the quasi-statistician who is actually aware of the assumption. Meanwhile, well-trained statisticians often do not check this assumption, a habit developed by the uncritical use of statistics with the PC.

The professional statistician has also been empowered by the PC's computational strength; without it, the natural seven-step cycle of statistical analysis would not be practical. [1] The PC and the analytical cycle comprise the perfect pairing as long as the steps are followed in order and the information obtained from one step is used in the next step. Unfortunately, statisticians are human and succumb to taking shortcuts through the seven-step cycle. They ignore the cycle and focus solely on the sixth step listed below. However, careful statistical endeavor requires additional procedures, as described in the seven-step cycle[1] that follows:

1. *Definition of the problem:* Determining the best way to tackle the problem is not always obvious. Management objectives are often expressed qualitatively, in which case the selection of the outcome or target (dependent) variable is subjectively biased. When the objectives are clearly stated, the appropriate dependent variable is often not available, in which case a surrogate must be used.

2. *Determining technique:* The technique first selected is often the one with which the data analyst is most comfortable; it is not necessarily the best technique for solving the problem.

3. *Use of competing techniques:* Applying alternative techniques increases the odds that a thorough analysis is conducted.

4. *Rough comparisons of efficacy:* Comparing variability of results across techniques can suggest additional techniques or the deletion of alternative techniques.

5. *Comparison in terms of a precise (and thereby inadequate) criterion:* Explicit criterion is difficult to define; therefore, precise surrogates are often used.

6. *Optimization in terms of a precise, and similarly inadequate criterion:* Explicit criterion is difficult to define; therefore, precise surrogates are often used.

7. *Comparison in terms of several optimization criteria:* This constitutes the final step in determining the best solution.

---

[1] The seven steps are Tukey's. The annotations are the author's.

The founding fathers of classical statistics — Karl Pearson and Sir Ronald Fisher — would have delighted in the PC's ability to free them from time-consuming empirical validations of their concepts. Pearson, whose contributions include regression analysis, the correlation coefficient, the standard deviation (a term he coined), and the chi-square test of statistical significance, would have likely developed even more concepts with the free time afforded by the PC. One can further speculate that the PC's functionality would have allowed Fisher's methods of maximum likelihood estimation, hypothesis testing, and analysis of variance to have immediate, practical applications.

The PC took the classical statistics of Pearson and Fisher from their theoretical blackboards into the practical classroom and boardroom.[2,3] In the 1970s statisticians were starting to acknowledge that their methodologies had potential for wider applications. First, a computing device was required to perform any multivariate statistical analysis with an acceptable accuracy and within a reasonable time frame. Although techniques had been developed for a *small-data setting* consisting of one or two handfuls of variables and up to hundreds of records, the tabulation of data was computationally demanding, and almost insurmountable. With the inception of the microprocessor in the mid-1970s, statisticians immediately found their computing device in the PC. As the PC grew in its capacity to store bigger data and perform knotty calculations at greater speeds, statisticians started replacing hand-held calculators with desktop PCs in the classrooms. From the 1990s to the present, the PC offered statisticians advantages that were imponderable decades earlier.

1.2 Statistics and Data Analysis

As early as in 1957, Roy believed that the classical statistical analysis was to a large extent likely to be supplanted by assumption-free, nonparametric approaches, which were more realistic and meaningful. [2] It was an onerous task to understand the robustness of the classical (parametric) techniques to violations of the restrictive and unrealistic assumptions underlying their use. In practical applications, the primary assumption of "a random sample from a multivariate normal population" is virtually untenable. The effects of violating this assumption and additional model-specific assumptions (e.g., linearity between predictor and dependent variables, constant variance among errors, and uncorrelated errors) are difficult to determine with any exactitude. It is difficult to encourage the use of the statistical techniques, given that their limitations are not fully understood.

---

[2] Karl Person (1900s) contributions include regression analysis, the correlation coefficient, and the chi-square test of statistical significance. He coined the term 'standard deviation' in 1893.

[3] Sir Ronald Fisher (1920s) invented the methods of maximum likelihood estimation, hypothesis testing, and analysis of variance.

In 1962, in his influential paper *The Future of Statistics*, John Tukey expressed concern that the field of statistics was not advancing. [1] He felt there was too much focus on the mathematics of statistics and not enough on the analysis of data, and predicted a movement to unlock the rigidities that characterize the discipline. In an act of statistical heresy, Tukey took the first step toward revolutionizing statistics by referring to himself not as a statistician, but a data analyst. However, it was not until the publication of his seminal masterpiece *Exploratory Data Analysis* in 1977 when Tukey lead the discipline away from the rigors of statistical inference into a new area, known as EDA. [3] For his part, Tukey tried to advance EDA as a separate and distinct discipline from statistics, an idea that is not universally accepted today. EDA offered a fresh, assumption-free, nonparametric approach to problem solving, in which the analysis is guided by the data itself, and utilizes self-educating techniques, such as iteratively testing and modifying the analysis as the evaluation of feedback, to improve the final analysis for reliable results.

The essence of EDA is best described in Tukey's own words: "Exploratory data analysis is detective work — numerical detective work — or counting detective work — or graphical detective work... [It is] about looking at data to see what it seems to say. It concentrates on simple arithmetic and easy-to-draw pictures. It regards whatever appearances we have recognized as partial descriptions, and tries to look beneath them for new insights." [3, page 1.] EDA includes the following characteristics:

1. *Flexibility* — techniques with greater flexibility to delve into the data
2. *Practicality* — advice for procedures of analyzing data
3. *Innovation* — techniques for interpreting results
4. *Universality* — use all of statistics that apply to analyzing data
5. *Simplicity* – above all, the belief that simplicity is the golden rule

On a personal note, when I learned that Tukey preferred to be called a data analyst, I felt both validated and liberated because many of my own analyses fell outside the realm of the classical statistical framework. Furthermore, I had virtually eliminated the mathematical machinery, such as the calculus of maximum likelihood. In homage to Tukey, I will use the terms data analyst and data analysis rather than statistical analysis and statistician throughout the book.

1.3   EDA

Tukey's book is more than a collection of new and creative rules and operations; it defines EDA as a discipline that holds that data analysts fail only if they fail to try many things. It further espouses the belief that data analysts are especially successful if their detective work forces them to notice the unexpected. In other words, the philosophy of EDA is a trinity of *attitude* and *flexibility* to do whatever it takes to refine the analysis, and *sharp-sightedness* to observe the unexpected when it does appear. EDA is thus a self-propagating theory; each data analyst adds his or her own contribution, thereby contributing to the discipline, as I hope to accomplish with this book.

The sharp-sightedness of EDA warrants more attention, as it is a very important feature of the EDA approach. The data analyst should be a keen observer of those indicators that are capable of being dealt with success-fully, and use them to paint an analytical picture of the data. In addition to the ever-ready visual graphical displays as an *indicator* of what the data reveal, there are numerical indicators, such as counts, percentages, aver-ages and the other classical descriptive statistics (e.g., standard deviation, minimum, maximum and missing values). The data analyst's personal judgment and interpretation of indictors are not considered a bad thing, as the goal is to draw informal inferences, rather than those statistically significance inferences that are the hallmark of statistical formality.

In addition to visual and numerical indicators, there are the *indirect mes-sages* in the data that force the data analyst to take notice, prompting responses such as "the data look like...," or, "it appears to be... ." Indirect messages may be vague; but their importance is to help the data analyst draw informal inferences. Thus, indicators do not include any of the hard statistical apparatus, such as confidence limits, significance test, or standard errors.

With EDA, a new trend in statistics was born. Tukey and Mosteller quickly followed up in 1977 with the second EDA book (commonly referred to EDA II), *Data Analysis and Regression*, which recasts the basics of classical inferential procedures of data analysis and regression as an assumption-free, nonparametric approach guided by "(a) a sequence of philosophical attitudes... for effective data analysis, and (b) a flow of useful and adaptable techniques that make it possible to put these attitudes to work." [4, page vii.]

Hoaglin, Mosteller and Tukey in 1983 succeeded in advancing EDA with *Understanding Robust and Exploratory Analysis*, which provides an under-standing of how badly the classical methods behave when their restrictive assumptions do not hold, and offers alternative robust and exploratory meth-ods to broaden the effectiveness of statistical analysis. [5] It includes a col-lection of methods to cope with data in an informal way, guiding the identification of data structures relatively quickly and easily, and trading off optimization of objective for stability of results.

Hoaglin et al. in 1991 continued their fruitful EDA efforts with *Fundamen-tals of Exploratory Analysis of Variance*. [6] They recast the basics of the analysis of variance with the classical statistical apparatus (e.g., degrees of freedom, F ratios and p-values) in a host of numerical and graphical displays, which often give insight into the structure of the data, such as sizes effects, patterns and interaction and behavior of residuals.

EDA set off a burst of activity in the visual portrayal of data. Published in 1983, *Graphical Methods for Data Analysis* (Chambers et al.) presents new and old methods — some which require a computer, while others only paper and pencil — but all are powerful data analysis tools to learn more about data structure. [7] In 1986 du Toit et al. came out with *Graphical Exploratory Data Analysis,* providing a comprehensive, yet simple presentation of the topic. [8] Jacoby with *Statistical Graphics for Visualizing Univariate and Bivariate Data* (1997), and *Statistical Graphics for Visualizing Multivariate Data* (*1998*) carries out his objective to obtain pictorial representations of quantitative information by elucidating histograms, one-dimensional and enhanced scatterplots and nonparametric smoothing. [9,10] In addition, he successfully transfers graphical displays of multivariate data on a single sheet of paper, a two-dimensional space.

1.4   The EDA Paradigm

EDA presents a major paradigm shift in the ways models are built. With the mantra "Let your data be your guide," EDA offers a view that is a complete reversal of the classical principles that govern the usual steps of model building. The EDA declares the model must always follow the data, not the other way around, as in the classical approach.

In the classical approach, the problem is stated and formulated in terms of an outcome variable Y. It is assumed that the *true* model explaining all the variation in Y is known. Specifically, it is assumed that all the structures (predictor variables, X_i s) affecting Y and their forms are known and present in the model. For example, if Age effects Y, but the log of Age reflects the true relationship with Y, then log of Age must be present in the model. Once the model is specified, the data are taken through the model-specific analysis, which provides the results in terms of numerical values associated with the structures, or estimates of the true predictor variables' coefficients. Then, interpretation is made for declaring X_i an important predictor, assessing how Xi affects the prediction of Y, and ranking X_i in order of predictive importance.

Of course, the data analyst never knows the true model. So, familiarity with the content domain of the problem is used to explicitly put forth the true *surrogate* model, from which good predictions of Y can be made. According to Box, "all models are wrong, but some are useful." [11] In this case, the model selected provides serviceable predictions of Y. Regardless of the model used, the assumption of knowing the truth about Y sets the statistical logic in motion to cause likely bias in the analysis, results and interpretation.

In the EDA approach, not much is assumed beyond having some prior experience with content domain of the problem. The right attitude, flexibility and sharp-sightedness are the forces behind the data analyst, who assesses the problem and lets the data guide the analysis, which then suggests the structures and their forms of the model. If the model passes the validity check, then it is considered final and ready for results and interpretation to be made. If not, with the force still behind the data analyst, the analysis and/ or data are revisited until new structures produce a sound and validated model, after which final results and interpretation are made (see Figure 1.1). Without exposure to assumption violations, the EDA paradigm offers a degree of confidence that its prescribed exploratory efforts are not biased, at least in the manner of classical approach. Of course, no analysis is bias-free, as all analysts admit their own bias into the equation.

1.5 EDA Weaknesses

With all its strengths and determination, EDA as originally developed had two minor weaknesses that could have hindered its wide acceptance and great success. One is of a subjective or psychological nature, and the other is a misconceived notion. Data analysts know that failure to look into a multitude of possibilities can result in a flawed analysis, thus finding themselves in a competitive struggle against the data itself. Thus, EDA can foster in data analysts an insecurity that their work is never done. The PC can assist the data analysts in being thorough with their analytical due diligence, but bears no responsibility for the arrogance EDA engenders.

The belief that EDA, which was originally developed for the small-data setting, does not work as well with large samples is a misconception. Indeed, some of the graphical methods, such as the stem-and-leaf plots, and some of the numerical and counting methods, such as folding and binning, do breakdown with large samples. However, the majority of the EDA methodology is unaffected by data size. Neither the manner in which the methods are carried out, nor the reliability of the results is changed. In fact, some of the most powerful EDA techniques scale up quite nicely, but do require the PC to do the serious number crunching of the *big data*.[4] [12] For example, techniques such as ladder of powers, re-expressing and smoothing are very valuable tools for large sample or big data applications.

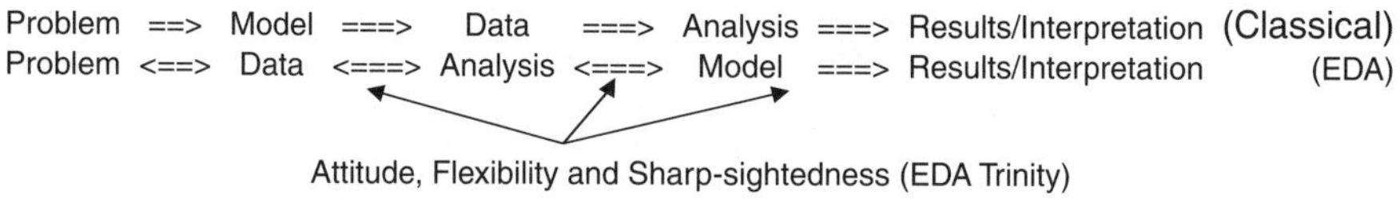

```
Problem  ==>  Model  ===>   Data    ===>  Analysis ===>  Results/Interpretation (Classical)
Problem  <==>  Data  <===>  Analysis <===>  Model   ===>  Results/Interpretation      (EDA)
```

Attitude, Flexibility and Sharp-sightedness (EDA Trinity)

FIGURE 1.1
EDA Paradigm

[4] Authors Weiss and Indurkhya and I use the general concept of "big" data. However, we stress different characteristics of the concept.

1.6   Small and Big Data

I would like to clarify the general concept of small and big data, as size, like beauty, is in the mind of the data analyst. In the past, small data fit the conceptual structure of classical statistics. Small always referred to the sample size, not the number of variables which were always kept to a handful. Depending on the method employed, small was seldom less than 5, was sometimes between 5 and 20, frequently between 30 and 50, and 50 and 100, and rarely between 100 and 200 individuals. In contrast to today's big data, small data are a tabular display of rows (observations or individuals) and columns (variables or features) that fits on a few sheets of paper.

In addition to the compact area they occupy, small data are neat and tidy. They are "clean," in that they contain no unexpected values, except for those due to primal data entry error. They do not include the statistical outliers and influential points, or the EDA far-out and outside points. They are in the "ready-to-run" condition required by classical statistical methods.

There are two sides to big data. On one side is classical statistics which considers big as simply being *not small*. Theoretically, big is the sample size after which asymptotic properties of the method "kick in" for valid results. On the other side is contemporary statistics which considers big in terms of *lifting observations* and *learning from the variables*. Although it depends on who is analyzing the data, a sample size greater than 50,000 individuals can be considered "big." Thus, calculating the average income from a database of 2 million individuals requires heavy-duty lifting (number crunching). In terms of learning or uncovering the structure among the variables, big can be considered 50 variables or more. Regardless of which side the data analyst is working, EDA scales up for both rows and columns of the data table.

1.6.1   Data Size Characteristics

There are three distinguishable characteristics of data size: *condition, location,* and *population.* Condition refers to the state of readiness of the data for analysis. Data that require minimal time and cost to clean before reliable analysis can be performed are well conditioned; data that involve a substantial amount of time and cost are ill conditioned. Small data are typically clean, and thus well conditioned.

Big data are an outgrowth of today's digital environment, which generates data flowing continuously from all directions at unprecedented speed and volume, and which almost always require cleansing. They are considered "dirty" mainly because of the merging of multiple sources. The merging process is inherently a time-intensive process, as multiple passes of the sources must be made to get a sense of how the combined sources fit together. Because of the iterative nature of the process, the logic of matching individual records across sources is at first "fuzzy," then fine-tuned to soundness; until that point unexplainable, seemingly random, nonsensical values result. Thus, big data are ill conditioned.

Location refers to where the data reside. Unlike the rectangular sheet for small data, big data reside in relational databases consisting of a *set of data tables*. The link among the data tables can be hierarchical (rank- or level-dependent) and/or sequential (time- or event-dependent). Merging of multiple data sources, each consisting of many rows and columns, produces data of even greater number of rows and columns, clearly suggesting bigness.

Population refers to the group of individuals having qualities or characteristics in common and related to the study under consideration. Small data ideally represent a random sample of a known population, which is not expected to encounter changes in its composition in the foreseeable future. The data are collected to answer a specific problem, permitting straightforward answers from a given problem-specific method. In contrast, big data often represent multiple, nonrandom samples of unknown populations, shifting in composition within the short-term. Big data are "secondary" in nature; that is, they are not collected for an intended purpose. They are available from the hydra of marketing information, for use on any post hoc query, and may not have a straightforward solution.

It is interesting to note that Tukey never talked specifically about the big data per se. However, he did predict that the cost of computing, both in time and dollars, would be cheap, which arguably suggests that he knew big data were coming. Regarding the cost, clearly today's PC bears this out.

1.6.2 Data Size: Personal Observation of One

The data size discussion raises the following question: how large should a sample be? Sample size can be anywhere from folds of 10,000 up to 100,000.

In my experience as a database marketing consultant and a teacher of statistics and data analysis, I have observed that the less experienced and trained statistician/data analyst uses sample sizes that are unnecessarily large. I see analyses performed on and models built from samples too large by factors ranging from 20 to 50. Although the PC can perform the heavy calculations, the extra time and cost in getting the larger data out of the data warehouse and then processing it and thinking about it are almost never justified. Of course, the only way a data analyst learns that extra big data are a waste of resources is by performing small vs. big data comparisons, a step I recommend.

1.7 Data Mining Paradigm

The term *data mining* emerged from the database marketing community sometime between the late 1970s and early 1980s. Statisticians did not understand the excitement and activity caused by this new technique, since the discovery of patterns and relationships (structure) in the data is not new to them. They had known about data mining for a long time, albeit under various names such as data fishing, snooping, and dredging, and most disparaging, "ransacking" the data. Because any discovery process inherently exploits the data, producing spurious findings, statisticians did not view data mining in a positive light.

Simply looking for something increases the odds that it will be found; therefore looking for structure typically results in finding structure. All data have spurious structures, which are formed by the "forces" that makes things come together, such as chance. The bigger the data, the greater are the odds that spurious structures abound. Thus, an expectation of data mining is that it produces structures, both real and spurious, without distinction between them.

Today, statisticians accept data mining only if it embodies the EDA paradigm. They define data mining as *any process that finds unexpected structures in data and uses the EDA framework to insure that the process explores the data, not exploits it* (see Figure 1.1). Note the word "unexpected," which suggests that the process is exploratory, rather than a confirmation that an expected structure has been found. By finding what one expects to find, there is no longer uncertainty as to the existence of the structure.

Statisticians are mindful of the inherent nature of data mining and try to make adjustments to minimize the number of spurious structures identified. In classical statistical analysis, statisticians have explicitly modified most analyses that search for interesting structure, such as adjusting the overall alpha level/type I error rate, or inflating the degrees of freedom [13,14]. In data mining, the statistician has no explicit analytical adjustments available, only the implicit adjustments affected by using the EDA paradigm itself. The following steps outline the data mining/EDA paradigm. As expected from EDA, the steps are defined by *soft* rules.

Suppose the objective is to find structure to help make good predictions of response to a future mail campaign. The following represent the steps that need to be taken:

Obtain the database that has similar mailings to the future mail campaign.

Draw of sample from the database. Size can be several folds of 10,000, up to 100,000.

Perform many exploratory passes of the sample. That is, do all desired calculations to determine the interesting or noticeable structure.

Stop the calculations that are used for finding the noticeable structure.

Count the number of noticeable structures that emerge. The structures are not final results and should not be declared significant findings.

Seek out indicators, visual and numerical, and the indirect messages.

React or respond to all indicators and indirect messages.

Ask questions. Does each structure make sense by itself? Do any of the structures form natural groups? Do the groups make sense; is there consistency among the structures within a group?

Try more techniques. Repeat the many exploratory passes with several fresh samples drawn from the database. Check for consistency across the multiple passes. If results do not behave in a similar way, there may be no structure to predict response to a future mailing, as chance may have infected your data. If results behave similarly, then assess the variability of each structure and each group.

Choose the most stable structures and groups of structures for predicting response to a future mailing.

1.8 Statistics and Machine Learning

Coined by Samuel in 1959, the term "machine learning" (ML) was given to the field of study that assigns computers the ability to learn without being explicitly programmed. [15] In other words, ML investigates ways in which the computer can acquire knowledge directly from data and thus learn to solve problems. It would not be long before machine learning would influence the statistical community.

In 1963, Morgan and Sonquist led a rebellion against the restrictive assumptions of classical statistics. [16] They developed the Automatic Interaction Detection (AID) regression tree, a methodology without assumptions. AID is a computer-intensive technique that finds or learns multidimensional patterns and relationships in data, and serves as an assumption-free, nonparametric alternative to regression analysis. AID marked the beginning of a nonstatistical machine learning approach to solving statistical problems. There have been many improvements and extensions of AID: THAID, MAID, CHAID, and CART, which are now considered valid data mining tools. CHAID and CART have emerged as the most popular today.

Independent from the work of Morgan and Sonquist, machine-learning researchers had been developing algorithms to automate the induction process, which provided another alternative to regression analysis. In 1979 Quinlan used the well-known Concept Learning System developed by Hunt et al. to implement one of the first intelligent systems — ID3 — which was succeeded by C4.5 and C5.0. [17,18] These algorithms are also considered data mining tools, but have not successfully crossed over to the statistical community.

The interface of statistics and machine learning began in earnest in the 1980s. Machine learning researchers became familiar with the three classical problems facing statisticians: regression (predicting a continuous outcome variable), classification (predicting a categorical outcome variable), and clustering (generating a few composite variables that carry a large percentage of the information in the original variables). They started using their machinery (algorithms and the PC) for a nonstatistical, or assumption-free, non-parametric approach to the three problem areas. At the same time, statisticians began harnessing the power of the desktop PC to influence the classical problems they know so well, thus relieving themselves from the starchy parametric road.

The machine-learning community has many specialty groups working on data mining: neural networks, fuzzy logic, genetic algorithms and programming, information retrieval, knowledge acquisition, text processing, inductive logic programming, expert systems, and dynamic programming. All areas have the same objective in mind, but accomplish it with their own tools and techniques. Unfortunately, the statistics community and the machine learning subgroups have no real exchanges of ideas or best practices. They create distinctions of no distinction.

1.9   Statistical Learning

In the spirit of EDA, it is incumbent on data analysts to try something new, and retry something old. They can benefit not only from the computational power of the PC in doing the heavy lifting of big data, but from the machine-learning ability of the PC in uncovering structure nestled in the big data. In the spirit of trying something old, statistics still has a lot to offer.

Thus, today's data mining can be defined in terms of three easy concepts:

1. *Statistics with emphasis on EDA proper:* This includes using the descriptive noninferential parts of classical statistical machinery as indicators. The parts include sum-of-squares, degrees of freedom, F ratios, chi-square values and p-values, but exclude inferential conclusions.

2. *Big data:* Big data are given special mention because of today's digital environment. However, because small data are a component of big data, it is not excluded.

3. *Machine-learning:* The PC is the learning machine, the *essential processing unit,* having the ability to learn without being explicitly programmed, and the intelligence to find structure in the data. Moreover, the PC is essential for big data, as it can always do what it is explicitly programmed to do.

In view of the above, the following data-mining mnemonic can be formed:

Data Mining = Statistics + Big data + Machine Learning and Lifting

Thus, data mining is defined today as *all of statistics and EDA for big and small data with the power of PC for the lifting of data and learning the structures within the data*. Explicitly referring to big and small data implies the process works equally well on both.

Again, in the spirit of EDA, it is prudent to parse the mnemonic equation. Lifting and learning require two different aspects of the data table and the PC. Lifting involves the rows of the data table and the capacity the PC in terms of MIPS (million instructions per second), the speed in which explicitly programmed steps are executed. An exception would be when there are many populations or clusters, in which case the PC will have to study the rows and their relationships to each other to identify the structure within the rows.

Learning focuses on the columns of the data table and the ability of the PC to find the structure within the columns without being explicitly programmed. Learning is more demanding on the PC than lifting, especially when the number of columns increases, in the same ways that learning from books is always more demanding than merely lifting the books. When lifting and learning of the rows are required in addition to learning within the columns, the PC must work exceptionally hard, but can yield extraordinary results.

Further parsing of the mnemonic equation reveals a branch of data mining referred to as *statistical learning*. Contracting the right-hand side of the mnemonic demonstrates that data mining *includes* statistical learning, given that the process makes no distinction between big and small data. *Statistical* connotes statistics and its classical problems of classification, prediction and clustering. *Learning* connotes the capacity of artificial or machine intelligence to exert its influence directly or indirectly on solving (statistical) problems. In other words, *statistical learning is solving statistical problems with nonstatistical machine learning*.

References

1. Tukey, J.W., The future of statistics, *Annals of Mathematical Statistics*, 33, 1–67, 1962.
2. Roy, S.N., *Some Aspects of Multivariate Analysis*, Wiley, New York, 1957.
3. Tukey, J.W., *The Exploratory Data Analysis*, Addison-Wesley, Reading, MA, 1977.
4. Mosteller, F. and Tukey, J.W., *Data Analysis and Regression*, Addison-Wesley, Reading, MA, 1977.
5. Hoaglin, D.C., Mosteller, F., and Tukey, J.W., *Understanding Robust and Exploratory Data Analysis*, John Wiley & Sons, New York, 1983.
6. Hoaglin, D.C., Mosteller, F., and Tukey, J.W., *Fundamentals of Exploratory Analysis of Variance*, John Wiley & Sons, New York, 1991.

7. Chambers, M.J., Cleveland, W.S., Kleiner, B., and Tukey, P.A., *Graphical Methods for Data Analysis*, Wadsworth & Brooks/Cole Publishing Company, CA, 1983.
8. du Toit, S.H.C., Steyn, A.G.W., and Stumpf, R.H., *Graphical Exploratory Data Analysis*, Springer-Verlag, New York, 1986.
9. Jacoby, W.G., *Statistical Graphics for Visualizing Univariate and Bivariate Data*, Sage Publication, Thousand Oaks, CA, 1997.
10. Jacoby, W.G., *Statistical Graphics for Visualizing Multivariate Data*, Sage Publication, Thousand Oaks, CA, 1998.
11. Box, G.E.P., Science and statistics, *Journal of the American Statistical Association*, 71, 791–799, 1976.
12. Weiss, S.M. and Indurkhya, N., *Predictive Data Mining*, Morgan Kaufman Publishers Inc., San Francisco, CA, 1998.
13. Dun, O.J., Multiple comparison among means, *Journal of the American Statistical Association*, 54, 52–64, 1961.
14. Ye, J., On measuring and correcting the effects of data mining and model selection, *Journal of the American Statistical Association*, 93, 120–131, 1998.
15. Samuel, A., Some studies in machine learning using the game of checkers, In Feigenbaum, E. and Feldman, J., Eds., *Computers and Thought*. McGraw-Hill, New York, 1963.
16. Morgan, J.N. and Sonquist, J.A., Problems in the analysis of survey data, and a proposal, *Journal of the American Statistical Association*, 58, 415–435, 1963.
17. Hunt, E., Marin, J., and Stone, P., *Experiments in Induction*, Academic Press, New York, 1966.
18. Quinlan, J.R., Discovering rules by induction from large collections of examples, In Mite, D., Ed., *Expert Systems in the Micro Electronic Age*, Edinburgh University Press, Edinburgh, U.K., 1979.

2

Two Simple Data Mining Methods for Variable Assessment

Assessing the relationship between a predictor variable and a target variable is an essential task in the model building process. If the relationship is identified and tractable, then the predictor variable is re-expressed to reflect the uncovered relationship, and consequently tested for inclusion into the model. Most methods of variable assessment are based on the well-known correlation coefficient, which is often misused because its linearity assumption is not tested. The purpose of this chapter is twofold: to present both the *smoothed scatterplot* as an easy and effective data mining method as well as another simple data mining method. The purpose of the latter is to assess a general association between two variables, while the purpose of the former is to embolden the data analyst to test the assumption to assure the proper use of the correlation coefficient.

I review the correlation coefficient with a quick tutorial, which includes an illustration of the importance of testing the linearity assumption, and outline the construction of the smoothed scatterplot, which serves as an easy method for testing the linearity assumption. Next, I present the *General Association Test*, a nonparametric test utilizing the smoothed scatterplot, as the proposed data mining method for assessing a general association between two variables.

2.1 Correlation Coefficient

The correlation coefficient, denoted by r, is a measure of the strength of the straight-line or linear relationship between two variables. The correlation coefficient takes on values ranging between +1 and –1. The following points are the accepted guidelines for interpreting the correlation coefficient:

1. 0 indicates no linear relationship.
2. +1 indicates a perfect positive linear relationship: as one variable increases in its values, the other variable also increases in its values via an exact linear rule.

3. –1 indicates a perfect negative linear relationship: as one variable increases in its values, the other variable decreases in its values via an exact linear rule.

4. Values between 0 and 0.3 (0 and –0.3) indicate a weak positive (negative) linear relationship via a shaky linear rule.

5. Values between 0.3 and 0.7 (0.3 and –0.7) indicate a moderate positive (negative) linear relationship via a fuzzy-firm linear rule.

6. Values between 0.7 and 1.0 (–0.7 and –1.0) indicate a strong positive (negative) linear relationship via a firm linear rule.

7. The value of r squared is typically taken as "the percent of variation in one variable explained by the other variable," or "the percent of variation shared between the two variables."

8. Linearity Assumption. The correlation coefficient requires that the underlying relationship between the two variables under consideration be linear. If the relationship is known to be linear, or the observed pattern between the two variables appears to be linear, then the correlation coefficient provides a reliable measure of the strength of the linear relationship. If the relationship is known to be nonlinear, or the observed pattern appears to be nonlinear, then the correlation coefficient is not useful, or at least questionable.

The calculation of the correlation coefficient for two variables, say X and Y, is simple to understand. Let zX and zY be the standardized versions of X and Y, respectively. That is, zX and zY are both re-expressed to have means equal to zero, and standard deviations (std) equal to one. The re-expressions used to obtain the standardized scores are in Equations (2.1) and (2.2):

$$zX_i = [X_i - \text{mean}(X)]/\text{std}(X) \qquad (2.1)$$

$$zY_i = [Y_i - \text{mean}(Y)]/\text{std}(Y) \qquad (2.2)$$

The correlation coefficient is defined as the mean product of the paired standardized scores (zX_i, zY_i) as expressed in Equation (2.3).

$$r_{X,Y} = \text{sum of } [zX_i * zY_i]/(n-1), \text{ where n is the sample size} \qquad (2.3)$$

For a simple illustration of the calculation, consider the sample of five observations in Table 2.1. Columns zX and zY contain the standardized scores of X and Y, respectively. The last column is the product of the paired standardized scores. The sum of these scores is 1.83. The mean of these scores (using the adjusted divisor n–1, not n) is 0.46. Thus, $r_{X,Y} = 0.46$.

TABLE 2.1

Calculation of Correlation Coefficient

obs	X	Y	zX	zY	zX*zY
1	12	77	−1.14	−0.96	1.11
2	15	98	−0.62	1.07	−0.66
3	17	75	−0.27	−1.16	0.32
4	23	93	0.76	0.58	0.44
5	26	92	1.28	0.48	0.62
mean	18.6	87		sum	1.83
std	5.77	10.32			
n	5			r	0.46

2.2 Scatterplots

The linearity assumption of the correlation coefficient can easily be tested with a *scatterplot*, which is a mapping of the paired points (X_i, Y_i) in a graph; i represents the observations from 1 to n, where n equals the sample size. For example, a picture that shows how two variables are related with respect to a horizontal/X-axis perpendicular to a vertical/Y-axis is representative of a scatterplot. If the scatter of points appears to overlay a straight-line, then the assumption has been satisfied, and $r_{X,Y}$ provides a meaningful measure of the linear relationship between X and Y. If the scatter does not appear to overlay a straight-line, then the assumption has not been satisfied and the $r_{X,Y}$ value is at best questionable. Thus, when using the correlation coefficient to measure the strength of the linear relationship, it is advisable to construct the scatterplot in order to test the linearity assumption. Unfortunately, many data analysts do not construct the scatterplot, thus rendering any analysis based on the correlation coefficient as potentially invalid. The following illustration is presented to reinforce the importance of evaluating scatterplots.

Consider the four data sets with eleven observations in Table 2.2. [1] There are four sets of (X,Y) points, with the same correlation coefficient value of 0.82. However, each X-Y relationship is distinct from one another, reflecting a different underlying structure as depicted in the scatterplots in Figure 2.1.

Scatterplot for X1-Y1 (upper-left) indicates a linear relationship; thus, the $r_{X1,Y1}$ value of 0.82 correctly suggests a strong positive linear relationship between X1 and Y1. Scatterplot for X2-Y2 (upper right) reveals a curved relationship; $r_{X2,Y2} = 0.82$. Scatterplot for X3-Y3 (lower left) reveals a straight-line except for the "outside" observation #3, data point (13, 12.74); $r_{X3,Y3} = 0.82$. Scatterplot for X4-Y4 (lower right) has a "shape of its own," which is clearly not linear; $r_{X4,Y4} = 0.82$. Accordingly, the correlation coefficient value of 0.82 is not a meaningful measure for the latter three X-Y relationships.

TABLE 2.2

Four Pairs of (X,Y) with the Same Correlation Coefficient (r = 0.82)

obs	X1	Y1	X2	Y2	X3	Y3	X4	Y4
1	10	8.04	10	9.14	10	7.46	8	6.58
2	8	6.95	8	8.14	8	6.77	8	5.76
3	13	7.58	13	8.74	13	12.74	8	7.71
4	9	8.81	9	8.77	9	7.11	8	8.84
5	11	8.33	11	9.26	11	7.81	8	8.47
6	14	9.96	14	8.1	14	8.84	8	7.04
7	6	7.24	6	6.13	6	6.08	8	5.25
8	4	4.26	4	3.1	4	5.39	19	12.5
9	12	10.84	12	9.13	12	8.15	8	5.56
10	7	4.82	7	7.26	7	6.42	8	7.91
11	5	5.68	5	4.74	5	5.73	8	6.89

2.3 Data Mining

Data mining — the process of revealing unexpected relationships in data — is needed to unmask the underlying relationships in scatterplots filled with big data. Big data are so much a part of the contemporary world that scatterplots have become overloaded with data points, or information. Ironically, scatterplots based on more information are actually less informative. With a quantitative target variable, the scatterplot typically becomes a cloud of points with sample-specific variation, called *rough*, which masks the underlying relationship. With a qualitative target variable, there is *discrete* rough, which masks the underlying relationship. In either case, if the rough can be removed from the big data scatterplot, then the underlying relationship can be revealed. After presenting two examples that illustrate how scatterplots filled with more data can actually provide less information, I outline the construction of the *smoothed scatterplot*, a rough-free scatterplot, which reveals the underlying relationship in big data.

2.3.1 Example #1

Consider the quantitative target variable TOLL CALLS (TC) in dollars, and the predictor variable HOUSEHOLD INCOME (HI) in dollars from a sample of size 102,000. The calculated $r_{TC, HI}$ is 0.09. The TC-HI scatterplot in Figure 2.2 shows a cloud of points obscuring the underlying relationship within the data (assuming a relationship exists). This scatterplot is not informative as to an indication for the reliable use of the calculated $r_{TC, HI}$.

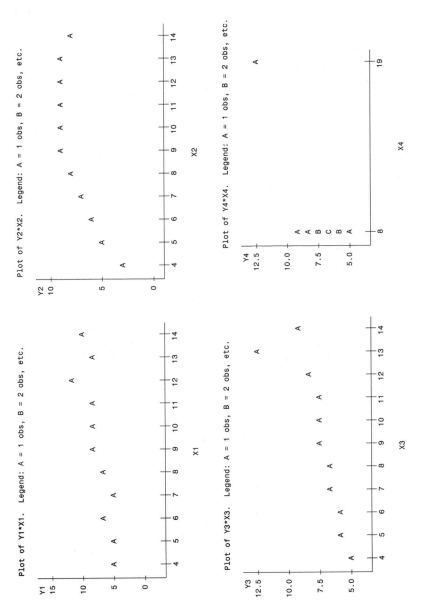

FIGURE 2.1
Four Different Datasets with the Same Correlation Coefficient

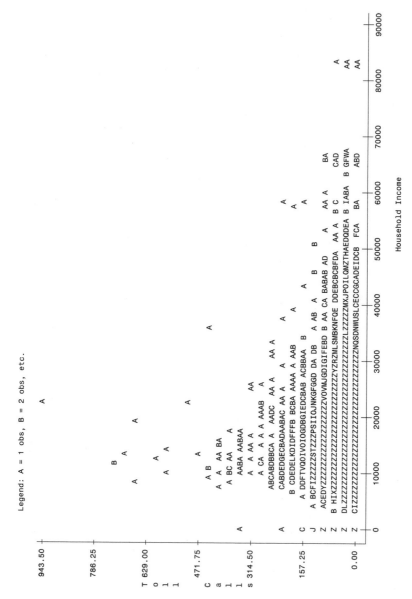

FIGURE 2.2

Scatterplot for TOLL CALLS and HOUSEHOLD INCOME

2.3.2 Example #2

Consider the qualitative target variable RESPONSE (RS), which measures the response to a mailing, and the predictor variable HOUSEHOLD INCOME (HI) from a sample of size 102,000. RS assumes "yes" and "no" values, which are coded as 1 and 0, respectively. The calculated $r_{RS, HI}$ is 0.01. The RS-HI scatterplot in Figure 2.3 shows "train tracks" obscuring the underlying relationship within the data (assuming a relationship exists). The tracks appear because the target variable takes on only two values, 0 and 1. As in the first example, this scatterplot is not informative as to an indication for the reliable use of the calculated $r_{RS, HI}$.

2.4   Smoothed Scatterplot

The *smoothed scatterplot* is the desired visual display for revealing a rough-free relationship lying within big data. *Smoothing* is a method of removing the rough and retaining the predictable underlying relationship (the *smooth*) in data by averaging within *neighborhoods* of similar values. Smoothing a X-Y scatterplot involves taking the averages of both the target (dependent) variable Y and the continuous predictor (independent) variable X, within X-based neighborhoods. [2] The six-step procedure to construct a smoothed scatterplot is as follows:

1. Plot the (X_i, Y_i) data points in a X-Y graph.
2. Slice the X-axis into several distinct and nonoverlapping neighborhoods or slices. A common approach to slicing the data is to create ten equal-sized slices (deciles), whose aggregate equals the total sample.[1] [3,4] (Smoothing with a categorical predictor variable is discussed in Chapter 3.)
3. Take the average of X within each slice. The mean or median can be used as the average. The average X values are known as smooth X values, or smooth decile X values.
4. Take the average of Y within each slice. For a quantitative Y variable, the mean or median can be used as the average. For a qualitative binary Y variable — which assumes only two values — clearly only the mean can be used. When Y is a 0–1 response variable, the average is conveniently a response rate. The average Y values are known as smooth Y values, or smooth decile Y values. For a qualitative Y variable, which assumes more than two values, the procedures discussed in Chapter 9 can be used.

---

[1] There are other designs for slicing continuous as well as categorical variables.

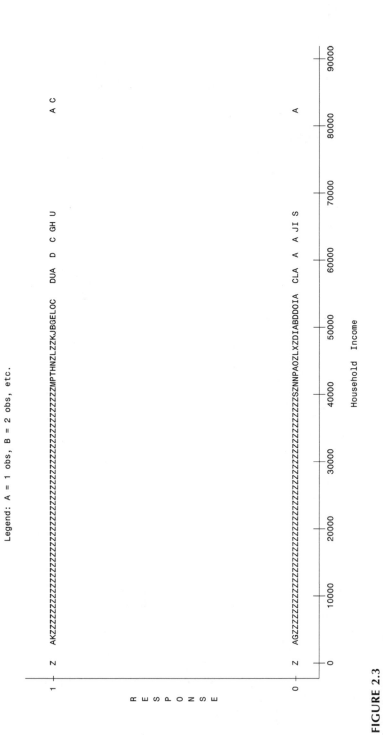

FIGURE 2.3

Scatterplot for RESPONSE and HOUSEHOLD INCOME

5. Plot the smooth points (smooth Y, smooth X), producing a *smoothed scatterplot*.

6. Connect the smooth points, starting from the left-most smooth point. The resultant *smooth trace* line reveals the underlying relationship between X and Y.

Returning to Examples #1 and #2, the HI data are grouped into ten equal-sized slices each consisting of 10,200 observations. The averages (means) or smooth points for HI with both TC and RS within the slices (numbered from 0 to 9) are presented in Tables 2.3 and 2.4, respectively. The smooth points are plotted and connected.

The TC smooth trace line in Figure 2.4 clearly indicates a linear relationship. Thus, the $r_{TC, HI}$ value of 0.09 is a reliable measure of a weak positive linear relationship between TC and HI. Moreover, the variable HI itself (without any re-expression) can be tested for inclusion in the TC Model. Note that the small r value does not preclude testing HI for model inclusion. This point is discussed in Chapter 4, Section 4.5.

The RS smooth trace line in Figure 2.5 is clear: the relationship between RS and HI is not linear. Thus, the $r_{RS, HI}$ value of 0.01 is invalid. This raises the following question: does the RS smooth trace line indicate a general association between RS and HI, implying a nonlinear relationship, or does the RS smooth trace line indicate a random scatter, implying no relationship between RS and HI? The answer can be readily found with the graphical nonparametric *General Association Test*. [5]

TABLE 2.3

Smooth Points: Toll Calls and Household Income

Slice	Average Toll Calls	Average HH Income
0	$31.98	$26,157
1	$27.95	$18,697
2	$26.94	$16,271
3	$25.47	$14,712
4	$25.04	$13,493
5	$25.30	$12,474
6	$24.43	$11,644
7	$24.84	$10,803
8	$23.79	$9,796
9	$22.86	$6,748

TABLE 2.4

Smooth Points: Response and Household Income

Slice	Average Response	Average HH Income
0	2.8%	$26,157
1	2.6%	$18,697
2	2.6%	$16,271
3	2.5%	$14,712
4	2.3%	$13,493
5	2.2%	$12,474
6	2.2%	$11,644
7	2.1%	$10,803
8	2.1%	$9,796
9	2.3%	$6,748

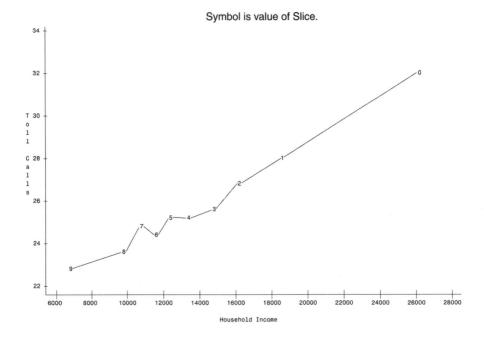

FIGURE 2.4

Smoothed Scatterplot for TOLL CALLS and HOUSEHOLD INCOME

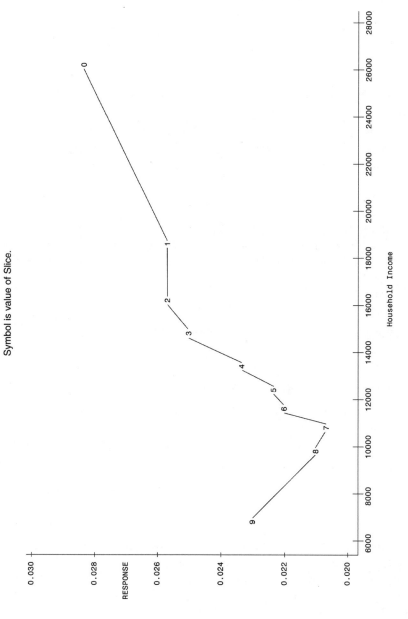

FIGURE 2.5

Smoothed Scatterplot for RESPONSE and HOUSEHOLD INCOME

2.5 General Association Test

Here is the General Association Test:

1. *Plot* the N smooth points in a scatterplot, and draw a horizontal medial line, which divides the N points into two equal-sized groups.

2. *Connect* the N smooth points starting from the left-most smooth point. N-1 line segments result. Count the number m of line segments that cross the medial line.

3. *Test* for significance. The null hypothesis is: there is no association between the two variables at hand. The alternative hypothesis is: there is an association between the two variables.

4. *Consider* the test-statistic: TS is $N - 1 - m$.

Reject the null hypothesis if TS is greater than or equal to the cutoff score in Table 2.5. It is concluded that there is an association between the two variables. The smooth trace line indicates the "shape" or structure of the association.

Fail-to-Reject the null hypothesis if TS is less than the cutoff score in Table 2.5. It is concluded that there is no association between the two variables.

Returning to the smoothed scatterplot of RS and HI, I determine the following:

1. There are ten smooth points: $N = 10$.

2. The medial line divides the smooth points such that points 5 to 9 are below the line and points 0 to 4 are above.

3. The line segment formed by points 4 and 5 in Figure 2.6 is the only segment that crosses the medial line. Accordingly, $m = 1$.

4. TS equals 8 ($= 10 - 1 - 1$), which is greater than/equal to the 95% and 99% confidence cutoff scores 7 and 8, respectively.

Thus, there is a 99% (and of course 95%) confidence level that there is an association between RS and HI. The RS smooth trace line in Figure 2.5 suggests that the observed relationship between RS and HI appears to be polynomial to the third power. Accordingly, the linear (HI), quadratic (HI^2) and cubed (HI^3) HOUSEHOLD INCOME terms should be tested in the RESPONSE model.

TABLE 2.5

Cutoff Scores for General Association Test (95% and 99% Confidence Levels)

N	95%	99%
8-9	6	–
10-11	7	8
12-13	9	10
14-15	10	11
16-17	11	12
18-19	12	14
20-21	14	15
22-23	15	16
24-25	16	17
26-27	17	19
28-29	18	20
30-31	19	21
32-33	21	22
34-35	22	24
36-37	23	25
38-39	24	26
40-41	25	27
42-43	26	28
44-45	27	30
46-47	29	31
48-49	30	32
50-51	31	33

2.6   Summary

It should be clear that an analysis based on the uncritical use of the coefficient correlation is problematic. The strength of a relationship between two variables cannot simply be taken as the calculated r value. The testing for the linearity assumption, which is made easy by the simple scatterplot or smoothed scatterplot, is necessary for a thorough and valid analysis. If the observed relationship is linear, then the r value can be taken at face value for the strength of the relationship at hand. If the observed relationship is not linear, then the r value must be disregarded, or used with extreme caution.

When a smoothed scatterplot for big data does not reveal a linear relationship, its scatter can be tested for randomness or for a noticeable general

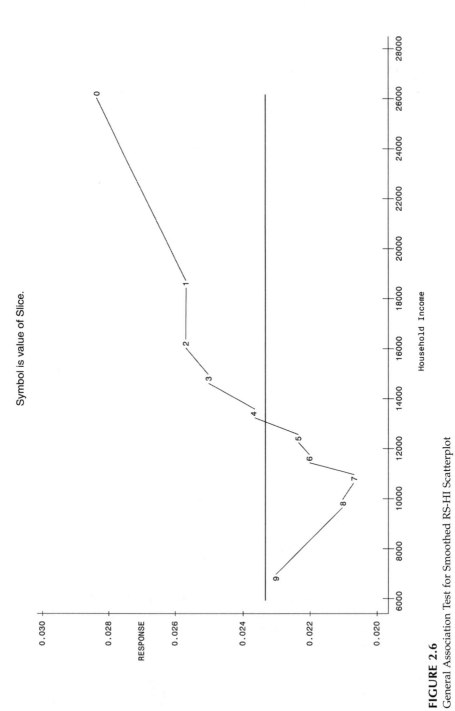

FIGURE 2.6
General Association Test for Smoothed RS-HI Scatterplot

association by the proposed nonparametric method. If the former is true, then it is concluded that there is no association between the variables. If the latter is true, then the predictor variable is re-expressed to reflect the observed relationship, and consequently tested for inclusion into the model.

References

1. Anscombe, F.J., Graphs in statistical analysis, *American Statistician*, 27, 17–22, 1973.
2. Tukey, J.W., *Exploratory Data Analysis*, Addison-Wesley, Reading, MA, 1997.
3. Hardle, W., *Smoothing Techniques*, Springer-Verlag, New York, 1990.
4. Simonoff, J.S., *Smoothing Methods in Statistics*, Springer-Verlag, New York, 1996.
5. Quenouille, M.H., *Rapid Statistical Calculations*, Hafner Publishing, New York, 1959.

3

Logistic Regression: The Workhorse of Database Response Modeling

Logistic regression is a popular technique for classifying individuals into two mutually exclusive and exhaustive categories, for example: buy–not buy or responder–nonresponder. It is the workhorse of response modeling as its results are considered the gold standard. Moreover, it is used as the benchmark for assessing the superiority of newer techniques, such as GenIQ, a genetic model, and older techniques, such as CHAID, which is a regression tree. In database marketing, response to a prior solicitation is the binary class variable (defined by responder and nonresponder), and the logistic regression model is built to classify an individual as either most likely or least likely to respond to a future solicitation.

In order to explain logistic regression, I first provide a brief overview of the technique and include the program code for the widely used SAS® (SAS Institute Inc., Cary, NC, 2002) system for building and scoring a logistic regression model. The code is a welcome addition to the techniques used by data analysts working on the two-group classification problem. Next, I present a case study demonstrating the building of a response model for an investment product solicitation. The model illustrates a host of data mining techniques that include the following:

Logit plotting

Re-expressing variables with the ladder of powers and the bulging rule

Measuring the straightness of data

Assessing the importance of individual predictor variables

Assessing the importance of a subset of predictor variables

Comparing the importance between two subsets of predictor variables

Assessing the relative importance of individual predictor variables

Selecting the best subset of predictor variables

Assessing goodness of model predictions

Smoothing a categorical variable for model inclusion

These techniques are basic skills that data analysts need to acquire; they are easy to understand, execute, and interpret, and should be mastered by anyone who wants control of the data and his or her findings.

3.1 Logistic Regression Model

Let Y be a binary class-dependent variable that assumes two outcomes or classes (typically labeled 0 and 1). The logistic regression model (LRM) classifies an individual into one of the classes based on the values for predictor (independent) variables X_1, X_2,..., X_n for that individual.

LRM estimates the *logit of Y* — a log of the odds of an individual belonging to class 1; the logit is defined in Equation (3.1). The logit, which takes on values between –7 and +7, is a virtually abstract measure for all but the experienced data analyst. (The logit theoretically assumes values between plus and minus infinity. However, in practice, it rarely goes outside the range of plus and minus seven.) Fortunately, the logit can easily be converted into the probability of an individual belonging to class 1, Prob(Y = 1), which is defined in Equation (3.2).

$$\text{logit } Y = b_0 + b_1{}^*X_1 + b_2{}^*X_2 + ... + b_n{}^*X_n \tag{3.1}$$

$$\text{Prob}(Y = 1) = \frac{\exp(\text{Logit } Y)}{1 + \exp(\text{Logit } Y)} \tag{3.2}$$

An individual's estimated (predicted) probability of belonging to class 1 is calculated by "plugging-in" the values of the predictor variables for that individual in the Equations (3.1) and (3.2). The bs are the logistic regression coefficients, which are determined by the calculus-based method of maximum likelihood. Note: Unlike the other coefficients, b_0 (referred to as the Intercept) has no predictor variable with which it is multiplied.

As presented, the LRM is readily and easily seen as the workhorse of database marketing response modeling, as the Yes-No response variable is an exemplary binary class variable. The illustration in the next section shows the rudimentaries of building a logistic regression response modeling.

3.1.1 Illustration

Consider dataset A, which consists of ten individuals and three variables in Table 3.1: the binary class variable RESPONSE (Y), INCOME in thousand-dollars (X1) and AGE in years (X2). I perform a logistic analysis regressing RESPONSE on INCOME and AGE using dataset A.

TABLE 3.1

Dataset A

Response (1=yes; 0=no)	Income ($000)	Age (years)
1	96	22
1	86	33
1	64	55
1	60	47
1	26	27
0	98	48
0	62	23
0	54	48
0	38	24
0	26	42

TABLE 3.2

LRM Output

Variable	df	Parameter Estimate	Standard Error	Wald Chi-Square	Pr> Chi-Square
INTERCEPT	1	−0.9367	2.5737	0.1325	0.7159
INCOME	1	0.0179	0.0265	0.4570	0.4990
AGE	1	−0.0042	0.0547	0.0059	0.9389

The standard LRM output in Table 3.2 includes the logistic regression coefficients and other "columns" of information (a discussion of these is beyond the scope of this chapter). The Estimate column contains the coefficients for INCOME, AGE and the INTERCEPT variables. The INTERCEPT variable is a mathematical device; it is implicitly defined as X_0, which is always equal to one (i.e., INTERCEPT = X_0 = 1). The coefficient b_0 is used as a "start" value given to all individuals regardless of their specific values for predictor variables in the model.

The estimated LRM is defined by Equation (3.3):

$$\text{logit RESPONSE} = -0.9367 + 0.0179 * \text{INCOME} - 0.0042 * \text{AGE} \quad (3.3)$$

Do not forget that LRM predicts the logit of RESPONSE, not the probability of RESPONSE.

3.1.2 Scoring a LRM

The SAS-code program in Figure 3.1 produces the LRM built with dataset A and scores an external dataset B in Table 3.3. The SAS procedure LOGISTIC produces logistic regression coefficients and puts them in the "coeff" file, as indicated by the code "outest = coeff." The coeff files produced by SAS versions 6 and 8 (SAS6, SAS8) are in Tables 3.4 and 3.5, respectively. The coeff files differ in two ways:

```
/****** Building the LRM on dataset A ***********/
PROC LOGISTIC data = A nosimple des outest = coeff;
model Response =
Income Age;
run;

/****** Scoring the LRM on dataset B ***********/
PROC SCORE data = B predict type = parms score = coeff
out= B_scored;
var Income Age;
run;

/******* Converting Logits into Probabilities ********/
        SAS version 6
data B_scored;
set B_scored;
Prob_Resp = exp(Estimate)/(1 + exp(Estimate));
run;
        SAS version 8
data B_scored;
set B_scored;
Prob_Resp = exp(Response)/(1 + exp(Response));
run;
```

FIGURE 3.1
SAS Code for Building and Score LRM

TABLE 3.3

Dataset B

Income ($000)	Age (years)
148	37
141	43
97	70
90	62
49	42

1. An additional column _STATUS_ in the SAS8 coeff file, which does not affect the scoring of the model, is this column.
2. The naming of the predicted logit is "Response" in SAS8, which is indicated by _NAME_ = Response.

Although it is unexpected, the naming of the predicted logit in SAS8 is the class variable used in the PROC LOGISTIC statement, as indicated by the code "model Response = ." In this illustration the predicted logit is called "Response," which is indicated by _NAME_ = Response, in Table 3.4. The SAS8 naming convention is unfortunate, as it may cause the analyst to think that the Response variable is a binary class variable and not a logit.

SAS procedure SCORE scores the five individuals in dataset B using the LRM coefficients, as indicated by the code "score = coeff." This effectively

TABLE 3.4

Coeff File (SAS6)

OBS	_LINK_	_TYPE_	_NAME_	Intercept	Income	Age	_LNLIKE_
1	LOGIT	PARMS	Estimate	-0.93671	0.017915	-0.0041991	-6.69218

TABLE 3.5

Coeff File (SAS8)

OBS	_LINK_	_TYPE_	_STATUS_	_NAME_	Intercept	Income	Age	_LNLIKE_
1	LOGIT	PARMS	0 Converged	Response	-0.93671	0.017915	-0.0041991	-6.69218

TABLE 3.6

Dataset B_scored

Income ($000)	Age (years)	Predicted Logit of Response: Estimate (SAS6), Response (SAS8)	Predicted Probability of Response: Prob_Resp
148	37	1.55930	0.82625
141	43	1.40870	0.80356
97	70	0.50708	0.62412
90	62	0.41527	0.60235
49	42	–0.23525	0.44146

appends the predicted logit variable (called Estimate when using SAS6 and Response when using SAS8), to the output file B_scored, as indicated by the code "out = B_scored," in Table 3.6. The probability of response (Prob_Resp) is easily obtained with the code at the end of the SAS-code program in Figure 3.1

3.2 Case Study

By examining the following case study about building a response model for a solicitation for investment products, I can illustrate a host of data mining techniques. To make the discussion of the techniques manageable, I use small data (a handful of variables, some of which take on few values, and a sample of size "petite grande") drawn from the original direct mail solicitation database. Reported results replicate the original findings obtained with slightly bigger data.

I allude to the issue of data size here in anticipation of data analysts who subscribe to the idea that big data are better for analysis and modeling. Currently there is a trend, especially in related statistical communities such as computer science, knowledge discovery and web mining, to use extra big data based on the notion that bigger is better. There is a statistical factoid that states if the true model can be built with small data, then the model built with extra big data produces large prediction error variance. Data analysts are never aware of the true model, but are guided when building it by the principle of simplicity. Therefore, it is wisest to build the model with small data. If the predictions are good, then the model is a good approximation of the true model; if predictions are not acceptable, then the EDA procedure prescribes an increase data size (by adding predictor variables and individuals) until the model produces good predictions. The data size, with which the model produces good predictions, is big enough. If extra big data are used, unnecessary variables tend to creep into the model, thereby increasing the prediction-error variance.

3.2.1　Candidate Predictor and Dependent Variables

Let TXN_ADD be the Yes-No response-dependent variable which records the activity of existing customers who received a mailing intended to motivate them to purchase additional investment products. Yes-No response, which is coded 1–0, respectively, corresponds to customers who have/have not added at least one new product fund to their investment portfolio. The TXN_ADD response rate is 11.9%, which is typically large for a direct mail campaign, but not unusual for solicitations intended to stimulate purchases among existing customers.

The five candidate predictor variables for predicting TXN_ADD whose values reflect measurement prior to the mailing are:

1. FD1_OPEN reflects the number of different types of accounts the customer has.
2. FD2_OPEN reflects the number of total accounts the customer has.
3. INVESTMENT reflects the customer's investment dollars in ordinal values: 1 = $25 – $499, 2 = $500 – $999, 3 = $1000 – $2999, 4 = $3000 – $4999, 5 = $5000 – $9999, and 6 = $10,000 +.
4. MOS_OPEN reflects the number of months the account is opened in ordinal values: 1 = 0 – 6 months, 2 = 7 – 12 months, 3 = 13 – 18 months, 4 = 19 – 24 months, 5 = 25 – 36 months, and 6 = 37 months+.
5. FD_TYPE is the product type of the customer's most recent investment purchase: A, B, C, ..., N.

3.3　Logits and Logit Plots

The logistic regression model belongs to the family of linear models that advance the implied critical assumption that the underlying relationship between a given predictor variable and the logit is linear or straight-line.

Bear in mind that to data analysts, the adjective "linear" refers to the explicit fact that the logit is expressed as the sum of weighted predictor variables, where the weights are the regression coefficients. In practice, however, the term refers to the implied assumption. To check this assumption, the *logit plot* is needed. A logit plot is the plot of the binary-dependent variable (hereafter, response variable) against the values of the predictor variable. Three steps are required to generate the logit plot:

1. Calculate the mean of the response variable corresponding to each value of the predictor variable. If the predictor variable takes on more than ten distinct values, then use *typical* values such as smooth decile values as defined in Chapter 2.
2. Calculate the logit of response using the formula that converts the mean of response to logit of response: logit = ln (mean/(1 – mean)), where ln is the natural logarithm.
3. Plot the logit of response values against the original distinct or the smooth decile values of the predictor variable.

One point worth noting is that the logit plot is an aggregate-level, not individual-level plot. The logit is an aggregate measure based on the mean of individual response values. Moreover, if smooth decile values are used, the plot is further aggregated, as each decile value represents 10% of the sample.

3.3.1   Logits for Case Study

For the case study, the response variable is TXN_ADD, and the logit of TXN_ADD is named LGT_TXN. For no particular reason, I start with candidate predictor variable FD1_OPEN, which takes on the distinct values 1, 2 and 3, in Table 3.7. Following the three-step construction for each FD1_OPEN value, I generate the LGT_TXN logit plot in Figure 3.2. I calculate the mean of TXN_ADD, and use the mean-to-logit conversion formula. For example, for FD1_OPEN = 1, the mean of TXN_ADD is 0.07, and the logit LGT_TXN is –2.4 (= ln(0.07/(1 – 0.07)). Last, I plot the LGT_TXN logit values against the FD1_OPEN values.

The LGT_TXN logit plot for FD1_OPEN does not suggest an underlying straight-line relationship between LGT_TXN and FD1_OPEN. In order to correctly use the logistic regression model, I need to straighten the relationship. A very effective and simple technique for straightening data is *re-expressing the variables*, which uses Tukey's Ladder of Powers and the Bulging Rule. Before presenting the details of the technique, it is worth discussing the importance of straight-line relationships or *straight data*.

TABLE 3.7

FD1_OPEN

FD1_OPEN	mean TXN_ADD	LGT_TXN
1	0.07	−2.4
2	0.18	−1.5
3	0.20	−1.4

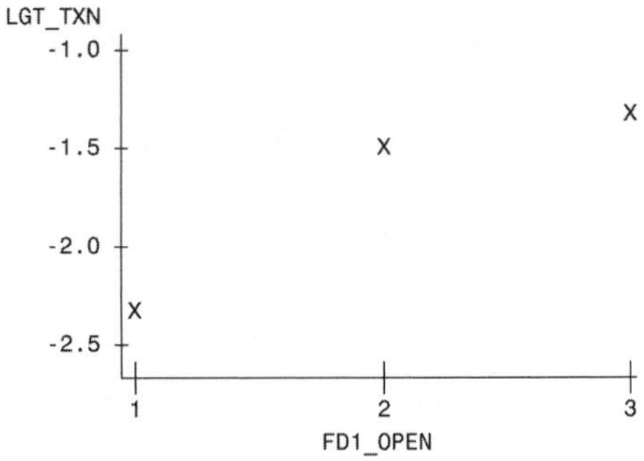

FIGURE 3.2
Logit Plot for FD1_OPEN

3.4   The Importance of Straight Data

EDA places special importance on straight data, not in the least for the sake of simplicity itself. The paradigm of life is simplicity (at least for those of us who are older and wiser). In the physical world, Einstein uncovered one of life's ruling principles using only three letters – $E = mc^2$. In the visual world, however, simplicity is undervalued and overlooked. A smiley face is an unsophisticated, simple shape that nevertheless communicates effectively, clearly and efficiently. Why, then, should the data analyst accept anything less than simplicity in his or her life's work? Numbers, as well, should communicate clearly, effectively and immediately. In the data analyst's world, there are two features that reflect simplicity — symmetry and straightness in the data. The data analyst should insist that the numbers be symmetric and straight.

The straight-line relationship between two continuous variables, say X and Y, is as simple as it gets. As X increases or decreases in its values, so does Y increase or decrease in its values, in which case it is said that X and Y are positively or negatively correlated, respectively. Or, as X increases (decreases) in its values, so does Y decrease (increase) in its values, in which case it is said that X and Y are negatively correlated. As further demonstration of its simplicity, Einstein's E and m have a perfect positively correlated straight-line relationship.

The second reason for the importance of straight data is that most database models require it, as they belong to the class of innumerable varieties of the linear model. Moreover, it has been shown that nonlinear models, which pride themselves on making better predictions with nonstraight data, in fact, do better with straight data.

I have not ignored the feature of symmetry. Not accidentally, as there are theoretical reasons, symmetry and straightness go hand-in-hand. Straightening data often makes data symmetric, and vice versa. You may recall that symmetric data often have the profile of the bell-shaped curve.

3.5 Re-expressing for Straight Data

Ladder of Powers is a method of re-expressing variables to straighten a bulging relationship between two continuous variables, say X and Y. Bulges in the data can be depicted as one of four shapes, as displayed in Figure 3.3. When the X-Y relationship has a bulge similar to any one of the four shapes, both the Ladder of Powers and the Bulging Rule, which guides the choice of "rung" in the ladder, are used to straighten out the bulge. Most data have bulges. However, when kinks or elbows characterize the data, then another approach is required, which will be discussed later in the chapter.

3.5.1 Ladder of Powers

Going up-ladder of powers means re-expressing a variable by raising it to a power p greater than 1. (Remember that a variable raised to the power of

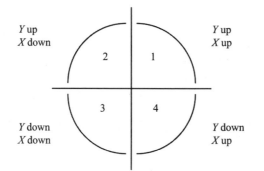

FIGURE 3.3
The Bulging Rule

1 is still that variable; $X^1 = X$, and $Y^1 = Y$). The most common p values used are 2 and 3. Sometimes values higher up-ladder and in-between values like 1.33 are used. Accordingly, starting at p = 1, the data analyst goes up-ladder, resulting in re-expressed variables, for X and Y, as follows:

$$\text{Starting at } X^1\text{: } X^2\ X^3\ X^4\ X^5...$$

$$\text{Starting at } Y^1\text{: } Y^2\ Y^3\ Y^4\ Y^5...$$

Some variables re-expressed going up-ladder have special names. Corresponding to power values 2 and 3, they are called X squared and X cubed, respectively, similarly for the Y variables.

Going down-ladder of powers means re-expressing a variable by raising it to a power p that is less than 1. The most common p values are $\frac{1}{2}$, 0, $-\frac{1}{2}$ and -1. Sometimes values lower down-ladder, and in-between values like 0.33 are used. Also, for negative powers, the re-expressed variable now sports a negative sign (i.e., is multiplied by -1), for which the reason is theoretical and beyond the scope of this chapter. Accordingly, starting at p = 1, the data analyst goes down-ladder, resulting in re-expressed variables, for X and Y, as follows:

$$\text{Starting at } X^1\text{: } X^{1/2}, X^0, -X^{-1/2}, -X^{-1},...$$

$$\text{Starting at } Y^1\text{: } Y^{1/2}, Y^0, -Y^{-1/2}, -Y^{-1},...$$

Some re-expressed variables down-ladder have special names. Corresponding to values, $\frac{1}{2} - \frac{1}{2}$ and -1, they are called square root of X, negative reciprocal square root of X, and negative reciprocal of X, respectively, similarly for the Y variables. The re-expression for p = 0 is not mathematically defined, and is conveniently defined as log to base 10. Thus $X^0 = \log X$, and $Y^0 = \log Y$.

3.5.2 Bulging Rule

The Bulging Rule states the following:

1. If the data have a shape similar to that shown in the first quadrant, then the data analyst tries re-expressing by going up-ladder for X, Y or both.
2. If the data have a shape similar to that shown in the second quadrant, then the data analyst tries re-expressing by going the down-ladder for X, and/or up-ladder for Y.

3. If the data have a shape similar to that shown in the third quadrant, then the data analyst tries re-expressing by going down-ladder for X, Y or both.

4. If the data have a shape similar to that shown in the fourth quadrant, then the data analyst tries re-expressing by going the up-ladder for X, and/or down-ladder for Y.

Re-expressing is an important, yet fallible part of EDA detective work. While it will typically result in straightening the data, it might result in a deterioration of information. Here is why: re-expression (going down too far) has the potential to squeeze the data so much that its values become indistinguishable, resulting in a loss of information. Expansion (going up too far) can potentially pull apart the data so much that the new far-apart values lie within an artificial range, resulting in a spurious gain of information.

Thus, re-expressing requires a careful balance between straightness and soundness. Data analysts can always go to the extremes of the ladder by exerting their will to obtain a little more straightness; but they must be mindful of a consequential loss of information. Sometimes it is evident when one has gone too far up/down on ladder, as there is power p, after which the relationship either does not improve noticeably, or inexplicably bulges in the opposite direction due to a corruption of information. I recommend using discretion to avoid over-straightening and its potential deterioration of information. Additionally, I caution that extreme re-expressions are sometimes due the extreme values of the original variables. Thus, always check the maximum and minimum values of the original variables to make sure they are reasonable before re-expressing the variables.

3.5.3 Measuring Straight Data

The correlation coefficient measures the strength of the straight-line or linear relationship between two variables X and Y and has been discussed in detail in Chapter 2. However, there is an additional assumption to consider.

In Chapter 2, I made reference to a "linear assumption," that the underlying relationship between X and Y is linear. The second assumption is an implicit one: that the (X, Y) data points are at the individual level. When the (X,Y) points are analyzed at an aggregate-level, such as in the logit plot and other plots presented in this chapter, the correlation coefficient based on "big" points tends to produce a "big" r value, which serves as a *gross estimate of the individual-level r value*. The aggregation of data diminishes the idiosyncrasies of the individual (X, Y) points, thereby increasing the resolution of the relationship, for which the r value also increases. Thus, the correlation coefficient on aggregated data serves as a gross indicator of the strength of the original X-Y relationship at hand. There is a drawback of aggregation: it often produces r values without noticeable differences because the power of the distinguishing individual-level information is lost.

3.6   Straight Data for Case Study

Returning to the LGT_TXN logit plot for FD1_OPEN, whose bulging relationship is in need of straightening, I identify its bulge as the type in quadrant
2 in Figure 3.3. According to the Bulging Rule, I should try going up-ladder
for LGT_TXN and/or down-ladder for FD1_OPEN. LGT_TXN cannot be re-
expressed because it is the explicit dependent variable as defined by the
logistic regression framework. Re-expressing it would produce grossly illogical results. Thus, I do not go up-ladder for LGT_TXN.

To go down-ladder for FD1_OPEN, I use the powers: $\frac{1}{2}$, 0, $-\frac{1}{2}$, –1 and
–2. This results in the square root of FD1_OPEN, labeled FD1_SQRT, the log
to base 10 of FD1_OPEN, labeled FD1_LOG, the negative reciprocal root of
FD1_OPEN, labeled FD1_RPRT, the negative reciprocal of FD1_OPEN,
labeled FD1_RCP, and the negative reciprocal square of FD1_OPEN, labeled
FD1_RSQ. The corresponding LGT_TXN logit plots for these re-expressed
variables and the original FD1_OPEN (repeated here for convenience) are
in Figure 3.4.

Visually, it appears that re-expressed variables FD1_RSQ, FD1_RCP and
FD1_RPRT do an equal job of straightening the data. I could choose any
one of them, but decide to do a little more detective work by looking at
the numerical indicator — the correlation coefficient between LGT_TXN
and the re-expressed variable in order to support my choice of the best re-
expressed variable. The larger the correlation coefficient, the more effective
the re-expressed variable is in straightening the data. Thus, the re-
expressed variable with the largest correlation coefficient is declared the
best re-expressed variable, with exceptions guided by the data analyst's
own experience with these visual and numerical indictors in the context
of the problem domain.

The correlation coefficients for LGT_ADD with FD1_OPEN, and with each
re-expressed variable are ranked in descending order in Table 3.8. The correlation coefficients of the re-expressed variables represent *noticeable*
improvements in straightening the data over the correlation coefficient for
the original variable FD1_OPEN (r = 0.907). FD1_RSQ has the largest correlation coefficient (r = 0.998), but it is slightly greater than that for FD1_RCP
(r = 0.988), and therefore not worthy of notice.

My choice of the best re-expressed variable is FD1_RCP, which represents
an 8.9% (= (0.988–0.907)/0.907) improvement in straightening the data over
the original relationship with FD1_OPEN. I prefer FD1_RCP over FD1_RSQ
and other extreme re-expressions down-ladder (defined by power p less than
–2) because I do not want to unwittingly select a re-expression that might
be too far down-ladder, resulting in loss of information. Thus, I go back one
rung to power –1, hoping to get the right balance between straightness and
minimal loss of information.

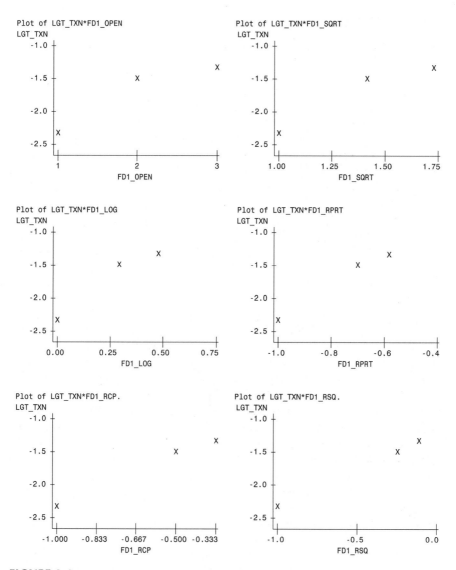

FIGURE 3.4

Logit Plots for FD1_OPEN and its Re-expressed Variables

TABLE 3.8

Correlation Coefficients between LGT_TXN and Re-expressed FD1_OPEN

FD1_RSQ	FD1_RCP	FD1_RPRT	FD1_LOG	FD1_SQRT	FD1_OPEN
0.998	0.988	0.997	0.960	0.937	0.907

TABLE 3.9

Correlation Coefficients between LGT_TXN and Re-expressed FD2_OPEN

FD2_RSQ	FD2_RCP	FD2_RPRT	FD2_LOG	FD2_SQRT	FD2_OPEN
0.995	0.982	0.968	0.949	0.923	0.891

3.6.1   Re-expressing FD2_OPEN

The scenario for FD2_OPEN is virtually identical to the one presented for FD1_OPEN, which is not surprising as FD1_OPEN and FD2_OPEN share a large amount of information. The correlation coefficient between the two variables is 0.97, meaning the two variables share 94.1% of their variation. Thus, I prefer FD2_RCP as the best re-expressed variable for FD2_OPEN. See Table 3.9.

3.6.2   Re-expressing INVESTMENT

The relationship between the LGT_TXN and INVESTMENT, depicted in the plot in Figure 3.10, is somewhat straight with a negative slope and a slight bulge in the middle for INVESTMENT values 3, 4 and 5. I identify the bulge of the type in quadrant 3 in Figure 3.3. Thus, I go down-ladder for powers $\frac{1}{2}$, 0, $-\frac{1}{2}$, -1, and -2, resulting in the square root of INVESTMENT, labeled INVEST_SQRT, the log to base 10 of INVESTMENT, labeled INVEST_LOG, the negative reciprocal root of INVESTMENT, labeled INVEST_RPRT, the negative reciprocal of INVESTMENT, labeled INVEST_RCP, and the negative reciprocal square of INVESTMENT, labeled INVEST_RSQ. The corresponding LGT_TXN logit plots for these re-expressed variables and the original INVESTMENT are in Figure 3.5.

Visually, I like the straight-line produced by INVEST_SQRT. Quantitatively, INVEST_LOG has the largest correlation coefficient, which supports the statistical factoid that claims a variable in dollar-units should be re-expressed with the log function. The correlation coefficients for INVEST_LOG and INVEST_SQRT in Table 3.10 are –0.978 and –0.966, respectively, which admittedly is not a noticeable difference. My choice of the best re-expression for INVESTMENT is INVEST_LOG because I prefer the statistical factoid over my visual choice. Only if a noticeable difference between correlation coefficients for INVEST_LOG and INVEST_SQRT existed would I sway from being guided by the factoid. INVEST_LOG represents an improvement of 3.4% (= (0.978 – 0.946)/0.946; disregarding the negative sign) in straightening the data over the relationship with the original variable INVESTMENT (r = –0.946).

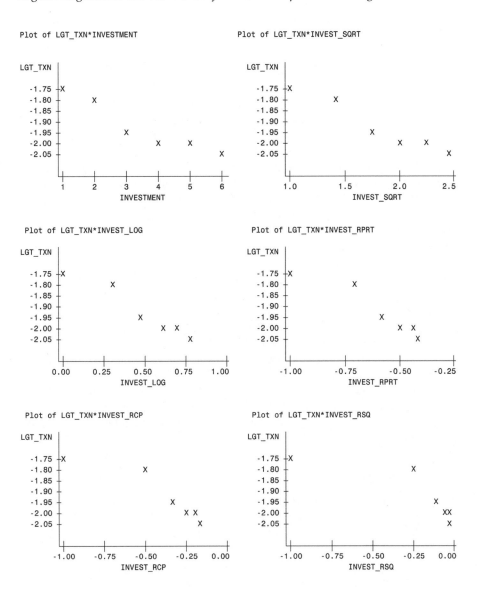

FIGURE 3.5
Logit Plots for INVESTMENT and its Re-expressed Variables

TABLE 3.10

Correlation Coefficients between LGT_TXN and Re-expressed INVESTMENT

INVEST_LOG	INVEST_SQRT	INVEST_RPRT	INVESTMENT	INVEST_RCP	INVEST_RSQ
−0.978	−0.966	−0.950	−0.946	−0.917	−0.840

3.7   Techniques When Bulging Rule Does Not Apply

I describe two plotting techniques for uncovering the correct re-expression when the Bulging Rule does not apply. After discussing the techniques, I return to the next variable for re-expression, MOS_OPEN. The relationship between LGT_TXN and MOS_OPEN is quite interesting, and offers an excellent opportunity to illustrate the flexibility of the EDA methodology.

It behooves the data analyst to perform due diligence to either qualitatively explain or quantitatively account for the relationship in a logit plot. Typically, the latter is easier than the former, as the data analyst is at best a scientist of data, not a psychologist of data. The data analyst seeks to investigate the plotted relationship to uncover the correct representation or *structure* of the given predictor variable. Briefly, *structure* is an organization of variables and functions. In this context, variables are broadly defined as both raw variables (e.g., X_1, X_2, ..., X_i, ...) and numerical constants, which can be thought as variables assuming any single value k, i.e., $X_i = k$. Functions include the arithmetic operators (addition, subtraction, multiplication and division), comparison operators (e.g., equal to, not equal, greater than), and logical operators (e.g., and, or, not, if ... then). For example, $X_1 + X_2/X_1$ is a structure.

By definition, any raw variable X_i is considered a structure, as it can be defined by $X_i = X_i+0$, or $X_i = X_i*1$. A dummy variable (X_dum) — a variable that assumes two numerical values, typically 1 and 0, which indicate the presence and absence of a condition, respectively — is structure. For example, X_dum = 1 if X equals 6; X_dum = 0 if X does not equal 6. The condition is "equals 6."

3.7.1   Fitted Logit Plot

The *fitted logit* plot is a valuable visual aid in uncovering and confirming structure. The fitted logit plot is defined as a plot of the *predicted* logit against a given structure. The steps required to construct the plot and its interpretation are:

1. *Perform* a logistic regression analysis on the response variable with the given structure, obtaining the predicted logit of response, as outlined in Section 3.1.2.

2. *Identify* the values of the structure to use in the plot. Identify the distinct values of the given structure. If the structure has more than ten values, identify its smooth decile values.

3. *Plot* the predicted (fitted) logit values against the identified-values of the structure. Label the points by the identified-values.

4. *Infer* that if the fitted logit plot reflects the shape in the original logit plot, the structure is the correct one. This further implies that the structure has some importance in predicting response. The extent to

which the fitted logit plot is different from original logit plot, the structure is a weak structure for predicting response.

3.7.2 Smooth Predicted vs. Actual Plot

Another valuable plot for exposing the detail of the strength or weakness of a structure is the *smooth predicted vs. actual* plot, which is defined as the plot of *mean* predicted response against *mean* actual response, for values of a reference variable. The steps required to construct the plot and its interpretation are as follows:

1. Calculate the mean predicted response by averaging the individual predicted probabilities of response from the appropriate logistic regression model, for each value of the reference variable. Similarly, calculate the mean actual response by averaging the individual actual responses, for each value of the reference variable.

2. The paired points (mean predicted response, mean actual response) are called *smooth points*.

3. Plot the smooth points, and label them by the values of the reference variable. If the reference variable has more than ten distinct values then use smooth decile values.

4. Insert the 45-degree line in the plot. The line serves as a reference for visual assessment of the importance of a structure for predicting response, thereby confirming that the structure under consideration is the correct one. Smooth points on the line imply that the mean predicted response and the mean actual response are equal, and there is great certainty that the structure is the correct one. The tighter the smooth points "hug" the 45-degree line, the greater the certainty of the structure. Conversely, the greater the scatter-about the line, the lesser the certainty of the structure.

3.8 Re-expressing MOS_OPEN

The relationship between LGT_TXN and MOS_OPEN in Figure 3.6 is not straight in the full range of MOS_OPEN values from one to six, but is straight between values one and five. The LGT_TXN logit plot for MOS_OPEN shows a check mark shape with vertex at MOS_OPEN = 5, as LGT_TXN jumps at MOS_OPEN = 6. Clearly, the Bulging Rule does not apply.

Accordingly, in uncovering the MOS_OPEN structure, I am looking for an organization of variables and functions that renders the ideal straight-line relationship between LGT_TXN and the MOS_OPEN structure. It will

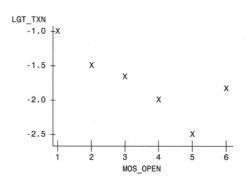

FIGURE 3.6
Logit Plot of MOS_OPEN

implicitly account for the jump in logit of TXN at MOS_OPEN = 6. Once the correct MOS_OPEN structure is identified, I can justifiably and explicitly include it in the TXN_ADD response model.

Exploring the structure of MOS_OPEN itself, I generate the LGT_TXN fitted logit plot, in Figure 3.7, which is based on the logistic regression analysis on TXN_ADD with MOS_OPEN. The logistic regression model, from which the predicted logits are obtained is defined in Equation (3.4):

$$logit(TXN\_ADD) = -1.24 - 0.17 * MOS\_OPEN \qquad (3.4)$$

It is acknowledged that MOS_OPEN has six distinct values. The fitted logit plot does not reflect the shape of the relationship in the original LGT_TXN

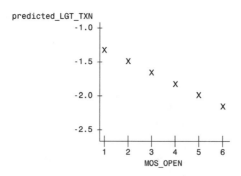

FIGURE 3.7
Fitted Logit Plot for MOS_OPEN

logit plot in Figure 3.6. The predicted point at MOS_OPEN = 6 is way too low. The implication and confirmation is that MOS_OPEN alone is not the correct structure as it does not produce the shape in the original logit plot.

3.8.1 Smooth Predicted vs. Actual Plot for MOS_OPEN

I generate the TXN_ADD smooth predicted vs. actual plot for MOS_OPEN, in Figure 3.8, which is both the structure under consideration and the reference variable. The smooth predicted values are based on the logistic regression model previously defined in Equation (3.4) and restated here in Equation (3.5) for convenience.

$$\text{logit(TXN\_ADD)} = -1.24 - 0.17 * \text{MOS\_OPEN} \tag{3.5}$$

There are six smooth points, each labeled by the corresponding six values of MOS_OPEN. The scatter-about the 45-degree line is wild, implying MOS_OPEN is not a good predictive structure, especially when MOS_OPEN equals 1, 5, 6, and 4, as their corresponding smooth points are not close to the 45-degree line. Point MOS_OPEN = 5 is understandable, as it can be considered the springboard to jump into LGT_TXN at MOS_OPEN = 6. MOS_OPEN = 1 as the farthest point from the line strikes me as inexplicable. MOS_OPEN = 4 may be within an acceptable distance from the line.

When MOS_OPEN equals 2 and 3, the prediction appears to be good, as the corresponding smooth points are close to the line. Two good predictions out of a possible six results in a poor 33% accuracy rate. Thus, MOS_OPEN is not good structure for predicting TXN_ADD. As before, the implication is that MOS_OPEN alone is not the correct structure to reflect the original relationship between LGT_TXN and MOS_OPEN in Figure 3.6. More detective work is needed.

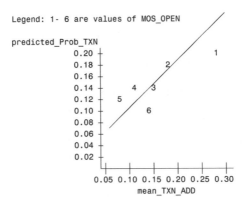

FIGURE 3.8
Plot of Smooth Predicted vs. Actual for MOS_OPEN

It occurs to me that the major problem with MOS_OPEN is the jump point. To explicitly account for the jump, I create a MOS_OPEN dummy variable structure defined as:

$$\text{MOS\_DUM} = 1 \text{ if MOS\_OPEN} = 6;$$

$$\text{MOS\_DUM} = 0 \text{ if MOS\_OPEN not equal to 6.}$$

I generate a second LGT_TXN fitted logit plot in Figure 3.9, but this time consisting of the predicted logits from regressing TXN_ADD on the structure consisting of MOS_OPEN and MOS_DUM. The logistic regression model is defined in Equation (3.6):

$$\text{logit (TXN\_ADD)} = -0.62 - 0.38*\text{MOS\_OPEN} + 1.16 *\text{MOS\_DUM} \quad (3.6)$$

This fitted plot accurately reflects the shape of the original relationship between TXN_ADD and MOS_OPEN in Figure 3.6. The implication is that MOS_OPEN and MOS_DUM make up the correct structure of the information carried in MOS_OPEN. The definition of the structure is the right side of the equation itself.

To complete my detective work, I create the second TXN_ADD smooth predicted vs. actual plot in Figure 3.10 consisting of mean predicted logits of TXN_ADD against mean MOS_OPEN. The predicted logits come from the logistic regression Equation (3.6) that includes the predictor variable pair MOS_OPEN and MOS_DUM. MOS_OPEN is used as the reference variable. The smooth points hug the 45-degree line nicely. The implication is that the MOS_OPEN-structure defined by MOS_OPEN and MOS_DUM is again confirmed, and the two-piece structure is an important predictive structure of TXN_ADD.

3.9 Assessing the Importance of Variables

The classic approach for assessing the statistical significance of a variable considered for model inclusion is the well-known null hypothesis–significance testing procedure, which is based on the reduction in prediction error (actual response minus predicted response) associated with the variable in question. The statistical apparatus of the formal testing procedure for logistic regression analysis consists of: the log likelihood function (LL), the G statistic, degrees of freedom, and the p-value. The procedure uses the apparatus within a theoretical framework with weighty and untenable assumptions. From a purist point of view, this could cast doubt on findings that actually have statistical significance. Even if findings of statistical significance are accepted as correct, they may not be of practical importance or have *noticeable* value

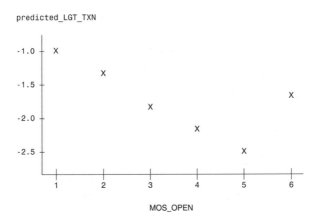

FIGURE 3.9
Fitted Logit Plot for MOS_OPEN and MOS_DUM

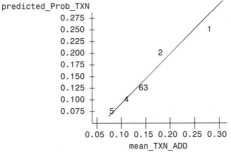

FIGURE 3.10
Plot of Smooth Predicted vs. Actual for MOS_OPEN and MOS_DUM

to the study at hand. For the data analyst with a pragmatic slant, the limitations and lack of scalability inherent in the classic system cannot be overlooked, especially within big data settings. In contrast, the data mining approach uses the LL units, the G statistic and degrees of freedom in an informal data-guided search for variables that suggest a *noticeable* reduction in prediction error. One point worth noting is that the informality of the data mining approach calls for suitable change in terminology, from declaring a result as statistically significant to one worthy of notice or *noticeably important*.

Before I describe the data mining approach of variable assessment, I would like to comment on the objectivity of the classic approach, as well as degrees of freedom. The classic approach is so ingrained in the analytic community, that no viable alternative occurs to practitioners, especially an alternative based on an informal and sometimes highly individualized series of steps. Declaring a variable statistically significant appears to be purely objective, as it is based on sound probability theory and statistical mathematical machinery. However, the settings of the testing machinery defined by analysts could affect the final results. The settings include the levels of rejecting a variable as significant, when, in fact, it is not, or accepting a variable as not significant, when, in fact, it is. Determining the proper sample size is also a subjective setting, as it depends on the amount budgeted for the study. Last, the allowable deviation of violations of test assumptions is set by the analyst's experience. Therefore, by acknowledging the subjective nature of the classic approach, the analyst can be receptive to the alternative data mining approach, which is free of theoretical ostentation and mathematical elegance.

A word about degrees of freedom will clarify the discussion. This concept is typically described as a generic measure of the number of independent pieces of information available for analysis. To insure accurate results, this concept is accompanied by the mathematical adjustment "replace N with N –1." The concept of degrees of freedom gives a deceptive impression of simplicity in counting the pieces of information. However, the principles used in counting are not easy, for all but the mathematical statistician. To date, there is no generalized calculus for counting degrees of freedom. Fortunately, the counting already exists for many analytical routines. Therefore, the correct degrees of freedom are readily available; computer output automatically provides them, and there are lookup tables in older statistics textbooks. For the analyses in the following discussions, the counting of degrees of freedom is provided.

3.9.1 Computing the G Statistic

In data mining, the assessment of the importance of a subset of variables for predicting response involves the notion of a *noticeable* reduction in prediction error due to the subset of variables, and is based on the ratio of the G statistic to the degrees of freedom (df), G/df. The degrees of freedom is defined as the number of variables in the subset. The G statistic is defined, in Equation (3.7) below, as the difference between two LL quantities, one corresponding to a model *without* the subset of variables, and the other corresponding to a model *with* the subset of variables.

G = –2 LL(model without variables) – –2 LL(model with variables)   (3.7)

There are two points worth noting: first, the LL units are multiplied by a factor of –2, a mathematical necessity; second, the term "subset" is used to imply that there is always a large set of variables available from which the analyst considers the smaller subset, which can include a single variable.

In the sections below, I detail the decision rules in three scenarios for assessing the likelihood that the variables have *some* predictive power. In brief, the larger the average G value per degrees of freedom (G/df), the more important the variables are in predicting response.

3.9.2 Importance of a Single Variable

If X is the only variable considered for inclusion into the model, the G statistic is defined in Equation (3.8)

$$G = -2 \text{ LL(model with Intercept only)} - -2 \text{ LL(model with X)} \quad (3.8)$$

The decision rule for declaring X an important variable in predicting response is: if G/df[1] is greater than the *standard G/df value 4*, then X is an important predictor variable and should be considered for inclusion in the model. Note that the decision rule only indicates that the variable has *some* importance, not *how much* importance. The decision rule implies that a variable with a greater G/df value has a greater *likelihood of some importance* than a variable with a smaller G/df value, not that it has greater importance.

3.9.3 Importance of a Subset of Variables

When subset A consisting of k variables is the only subset considered for model inclusion, the G statistic is defined in Equation (3.9):

$$G = -2 \text{ LL(model with Intercept)} - -2 \text{ LL(model with A(k) variables)} \quad (3.9)$$

The decision rule for declaring subset A important in predicting response is: if G/k is greater than the standard G/df value 4, then subset A is an important subset of predictor variable and should be considered for inclusion in the model. As before, the decision rule only indicates that the subset has some importance, not how much importance.

3.9.4 Comparing the Importance of Different Subsets of Variables

Let subsets A and B consist of k and p variables, respectively. The number of variables in each subset does not have to be equal. If they are equal, then all but one variable can be the same in both subsets. The G statistic for A and B are defined in Equations (3.10) and (3.11), respectively:

$$G(k) = -2 \text{ LL(model with Intercept)} - -2 \text{ LL(model with "A" variables)} \quad (3.10)$$

$$G(p) = -2 \text{ LL(model with Intercept)} - -2 \text{ LL(model with "B" variables)} \quad (3.11)$$

---

[1] Obviously, G/df equals G for single a predictor variable with df = 1.

The decision rule for declaring which of the two subsets is more important (greater likelihood of having some predictive power) in predicting response is as follows:

1. If G(k)/k greater than G(p)/p, then subset A is the more important predictor variable subset; otherwise, B is the more important subset.
2. If G(k)/k and G(p)/p are equal or have comparable values, then both subsets are to be regarded tentatively of comparable importance. The data analyst should consider additional indicators to assist in the decision about which subset is better.

It clearly follows from the decision rule that the better model is defined by the more important subset. Of course, this rule assumes that G(k)/k and G(p)/p are greater than the standard G/df value 4.

3.10 Important Variables for Case Study

The first step in variable assessment is to determine the baseline LL value for the data under study. The logistic regression model for TXN_ADD without variables produces two essential bits of information in Table 3.11:

1. The baseline for this case study is –2LL equals 3606.488
2. The logistic regression model is defined in Equation (3.12)

$$\text{Logit(TNX\_ADD} = 1) = -1.9965 \tag{3.12}$$

There are interesting bits of information in Table 3.11 that illustrate two useful statistical identities:

1. Exponentiation of both sides of Equation (3.12) produces the odds of response equals 0.1358. Recall that exponentiation is the mathematical operation of raising a quantity to a power. The exponentiation of a logit is the odds, and consequently, the exponentiation of –1.9965 is 0.1358.

$$\text{Exp(Logit(TNX\_ADD} = 1)) = \text{Exp}(-1.9965) \tag{3.13}$$

$$\text{Odds (TNX\_ADD} = 1) = \text{Exp}(-1.9965) \tag{3.14}$$

$$\text{Odds (TNX\_ADD} = 1) = 0.1358 \tag{3.15}$$

TABLE 3.11

The LOGISTIC Procedure for TXN_ADD

Response Profile		
TXN_ADD	COUNT	
1	589	
0	4337	
−2LL = 3606.488		

Variable	Parameter Estimate	Standard Error	Wald Chi-Square	Pr > Chi-Square
INTERCEPT	−1.9965	0.0439	2067.050	0.0

$$\text{LOGIT} = 1.9965$$
$$\text{ODDS} = \text{EXP}(.19965) = .1358$$

$$\text{PROB(TXN\_ADD} = 1) = \frac{\text{ODDS}}{1 + \text{ODDS}} = \frac{0.1358}{1 + 0.1358} = 0.119$$

2. Probability of (TNX_ADD = 1), hereafter, the probability of RESPONSE, is easily obtained as the ratio of odds divided by 1+odds. The implication is that the best estimate of RESPONSE — when no information is known or no variables are used — is 11.9%, namely, the average response of the mailing.

3.10.1 Importance of the Predictor Variables

With the LL baseline value 3606.488, I assess the importance of the five variables: MOS_OPEN and MOS_DUM, FD1_RCP, FD2_RCP and INVEST_LOG. Starting with MOS_OPEN and MOS_DUM, as they must be together in the model, I perform a logistic regression analysis on TXN_ADD with MOS_OPEN and MOS_DUM; the output is in Table 3.12. From Equation (3.9), the G value is 107.022 (= 3606.488 − 3499.466). The degrees of freedom is equal to the number of variables; df is 2. Accordingly, G/df equals 53.511, which is greater the standard G/df value 4. Thus MOS_OPEN and MOS_DUM as a pair are declared important predictor variables of TXN_ADD.

From Equation (3.8), the G/df value for each remaining variable in Table 3.12 is greater than 4. Thus, these five variables, each important predictors of TXN_ADD, form a *starter subset* for predicting TXN_ADD. I have not forgotten about FD_TYPE; it will be discussed later.

I build a preliminary model by regressing TXN_ADD on the starter subset; the output is in Table 3.13. From Equation (3.9), the five-variable subset has a G/df value of 40.21(= 201.031/5), which is greater than 4. Thus, this is an *incipient subset* of important variables for predicting TXN_ADD.

TABLE 3.12

G and df for Predictor Variables

Variable	–2LL	G	df	p-value
INTERCEPT	3606.488			
MOS_OPEN + MOS_DUM	3499.466	107.023	2	0.0001
FD1_RCP	3511.510	94.978	1	0.0001
FD2_RCP	3503.993	102.495	1	0.0001
INV_LOG	3601.881	4.607	1	0.0001

TABLE 3.13

Preliminary Logistic Model for TXN_ADD with Starter Subset

	Intercept Only	Intercept and All Variables	All Variables	
–2LL	3606.488	3405.457	201.031 with 5 df (p = 0.0001)	
Variable	*Parameter Estimate*	*Standard Error*	*Wald Chi-Square*	*Pr > Chi-Square*
INTERCEPT	0.9948	0.2462	16.3228	0.0001
FD2_RCP	3.6075	0.9679	13.8911	0.0002
MOS_OPEN	-0.3355	0.0383	76.8313	0.0001
MOS_DUM	0.9335	0.1332	49.0856	0.0001
INV_LOG	-0.7820	0.2291	11.6557	0.0006
FD1_RCP	-2.0269	0.9698	4.3686	0.0366

3.11 Relative Importance of the Variables

The mystery in building a statistical model is that the *true* subset of variables defining the *true* model is not known. The data analyst can be most productive by seeking to find the *best* subset of variables that defines the final model as an "intelli-guess" of the true model. The final model reflects more of the analyst's effort given the data at hand than an estimate of the true model itself. The analyst's attention has been drawn to the most noticeable, unavoidable collection of predictor variables, whose behavior is known to the extent the logit plots uncover their shapes and their relationships to response.

There is also magic in building a statistical model, in that the best subset of predictor variables consists of variables whose contributions to the model's predictions are often unpredictable and unexplainable. Sometimes the most important variable in the mix drops from the top, in that its contribution in the model is no longer as strong as it was individually. Other times the least unlikely variable rises from the bottom, in that its contribution in the model is stronger than it was individually. In the best of times, the variables interact with each other such that their total effect on the model's predictions is greater than the sum of their individual effects.

Unless the variables are not correlated with each other (the rarest of possibilities), it is impossible for the analyst to assess a variable's unique contribution. In practice, the analyst can assess a variable's *relative importance*, specifically, its importance with respect to the presence of the other variables in the model. The Wald chi-square — as posted in logistic regression analysis output — serves as an indicator of a variable's relative importance, as well as for selecting the best subset. This is discussed in the next section.

3.11.1 Selecting the Best Subset

The decision rules for finding the best subset of important variables consist of the following steps:

1. *Select an initial subset of important variables.* Variables that are thought to be important are probably important; let experience (yours and others) in the problem domain be the rule. If there are many variables from which to choose, rank the variables based on the correlation coefficient r (between response variable and candidate predictor variable). One to two handfuls of the experience-based variables, the largest r-valued variables, and some small r-valued variables form the initial subset. The latter variables are included because small r-values may falsely exclude important nonlinear variables. (Recall that the correlation coefficient is an indicator of linear relationship.) Categorical variables require special treatment, as the correlation coefficient cannot be calculated. (I illustrate with FD_TYPE how to include a categorical variable in a model in the last section.)

2. For the variables in the initial subset, *generate logit plots and straighten the variables as required.* The most noticeable handfuls of original and re-expressed variables form the starter subset.

3. *Perform the preliminary logistic regression analysis on the starter subset.* Delete one or two variables with Wald chi-square values less than the *Wald cut-off value 4* from the model. This results in the first incipient subset of important variables.

4. *Perform another logistic regression analysis on the incipient subset.* Delete one or two variables with Wald chi-square values less than the Wald cut-off value 4 from the model. The analyst can create an illusion of important variables appearing and disappearing with the deletion of different variables. The Wald chi-square values can exhibit "bouncing" above and below the Wald cut-off value as the variables are deleted. The bouncing effect is due to the correlation between the "included" variables and the "deleted" variables. The greater the correlation, the greater the bouncing (unreliability) of Wald chi-

square values, and consequently, the greater the uncertainty of declaring important variables.

5. *Repeat step 4 until all retained predictor variables have comparable Wald chi-square values.* This step often results in different subsets, as the data analyst deletes judicially different pairings of variables.

6. *Declare the best subset by comparing the relative importance of the different subsets using decision rule in Section 3.9.4.*

3.12 Best Subset of Variables for Case Study

I perform a logistic regression on TXN_ADD with the five-variable subset, MOS_OPEN and MOS_DUM, FD1_RCP, FD2_RCP and INVEST_LOG; the output is in Table 3.13. FD1_RCP has the smallest Wald Chi-square value 4.3686. FD2_RCP, which has a Wald chi-square 13.8911, is highly correlated with FD1_RCP ($r_{FD1\_RCP, FD2\_RCP}$ = 0.97), thus rendering their Wald chi-square values unreliable. However, without additional indicators for either variable, I accept their "face" values as an indirect message and delete FD1_RCP, the variable with the lesser value.

INVEST_LOG has the second smallest Wald chi-square value 11.6557. With no apparent reason other than it is just appears to have a less relative importance given MOS_OPEN, MOS_DUM, FD1_RCP and FD2_RCP in the model, I also delete INVEST_LOG from the model. Thus, the incipiently best subset consists of FD2_RCP, MOS_OPEN and MOS_DUM.

I perform another logistic regression on TXN_ADD with the three-variable subset (FD2_RCP, MOS_OPEN and MOS_DUM); the output is in Table 3.14. MOS_OPEN and FD2_RCP have comparable Wald chi-square values, 81.8072 and 85.7923, respectively, which are obviously greater than the Wald

TABLE 3.14

Logistic Model for TXN_ADD with Best Incipient Subset

	Intercept Only	Intercept and All Variables	All Variables	
–2LL	3606.488	3420.430	186.058 with 3 df (p = 0.0001)	
Variable	*Parameter Estimate*	*Standard Error*	*Wald Chi-Square*	*Pr > Chi-Square*
INTERCEPT	0.5164	0.1935	7.1254	0.0076
FD2_RCP	1.4942	0.1652	81.8072	0.0001
MOS_OPEN	–0.3507	0.0379	85.7923	0.0001
MOS_DUM	0.9249	0.1329	48.4654	0.0001

cut-off value 4. The Wald chi-square value for MOS_DUM is half of that MOS_OPEN, and not comparable to the other values. However, MOS_DUM is staying in the model because it is empirically needed (recall Figures 3.9 and 3.10). I acknowledge that MOS_DUM and MOS_OPEN share information, which could be effecting the reliability of their Wald chi-square values. The actual amount of shared information is 42%, which indicates there is a minimal effect on the reliability of their Wald chi-square values.

I compare the importance of the current three-variable subset (FD2_RCP, MOS_OPEN, MOS_DUM) and the starter five-variable subset (MOS_OPEN, MOS_DUM, FD1_RCP, FD2_RCP, INVEST_LOG). The G/df values are 62.02 (= 186.058/3 from Table 3.14) and 40.21 (= 201.031/5 from Table 3.13), for the former and latter subsets, respectively. Based on the decision rule in Section 3.9.4, I declare the three-variable subset is better than five-variable subset. Thus, I expect good predictions of TXN_ADD based on the three-variable model defined in Equation (3.16):

$$\text{Predicted logit of TXN\_ADD} =$$

$$\text{Predicted LGT\_TXN} =$$

$$0.54164. + 1.4942^*\text{FD2\_RCP} - 0.3507^*\text{MOS\_OPEN} + 0.9249^*\text{MOS\_DUM}$$

$$(3.16)$$

3.13 Visual Indicators of Goodness of Model Predictions

In this section I provide visual indicators of the quality of model predictions. The logistic regression model itself is a variable, as it is a sum of weighted variables with the logistic regression coefficients serving as the weights. As such, the logit model predictions (e.g., the predicted LGT_TXN) is a variable, which has a mean, a variance, and all the other descriptive measures afforded any variable. Also, the logit model predictions can be graphically displayed as afforded any variable. Accordingly, I present three valuable plotting techniques, which reflect the EDA-prescribed "graphic detective work" for assessing the goodness of model predictions.

3.13.1 Smooth Residual by Score Groups Plot

The *smooth residual by score groups plot* is defined as the plot consisting of the mean residual against the mean predicted response by *score groups*, which are identified by the unique values created by preselected variables, typically the predictor variables in the model under consideration. For example, for the three-variable model, there are eighteen score groups: three values of

FD2_RCP multiplied by six values of MOS_OPEN. The two values of MOS_DUM are not unique, as they are part of the values of MOS_OPEN.

The steps required to construct the smooth residual by score groups plot and its interpretation are as follows:

1. *Score* the data by appending the predicted logit as outlined in Section 3.1.2.
2. *Convert* the predicted logit to predicted probability of response as outlined in Section 3.1.2.
3. *Calculate* the residual (error) for each individual: residual = actual response minus predicted probability of response.
4. *Determine* the score groups by the unique values created by the pre-selected variables.
5. For each score group, *calculate* the mean (smooth) residual and mean (smooth) predicted response, producing a set of paired smooth points (smooth residual, smooth predicted response).
6. *Plot* the smooth points by score group.
7. *Draw* a straight line through mean residual = 0. This zero-line serves as a reference line for determining whether or not a general trend exists in the scatter of smooth points. If the smooth residual plot looks like the ideal or *null* plot, i.e., has a random scatter-about the zero-line with about half of the points above the line and the remaining points below, then it is concluded that there is no general trend in the smooth residuals. Thus, the predictions aggregated at the score-group level are considered good. The desired implication is that on average the predictions at the *individual-level* are also good.
8. *Examine* the smooth residual plot for noticeable deviations from random scatter. This is at best a subjective task, as it depends on the data analyst's unwitting nature to see what is desired. To aid in an objective examination of the smooth residual plot, use the General Association Test discussed in Chapter 2 to determine whether the smooth residual plot is equivalent to the null plot.
9. When the smooth residual plot is declared null, *look* for a local pattern. It is not unlikely for a small wave of smooth points to form a local pattern, which has no ripple effect to create a general trend in an otherwise null plot. A local pattern indicates a weakness or *weak spot* in the model, in that there is a prediction bias for the score groups identified by the pattern.

3.13.1.1 Smooth Residual by Score Groups Plot for Case Study

I construct the smooth residual by score groups plot to determine the quality of the predictions of the three-variable (FD2_RCP, MOS_OPEN, MOS_DUM) model. The smooth residual plot in Figure 3.11 is declared to be equivalent

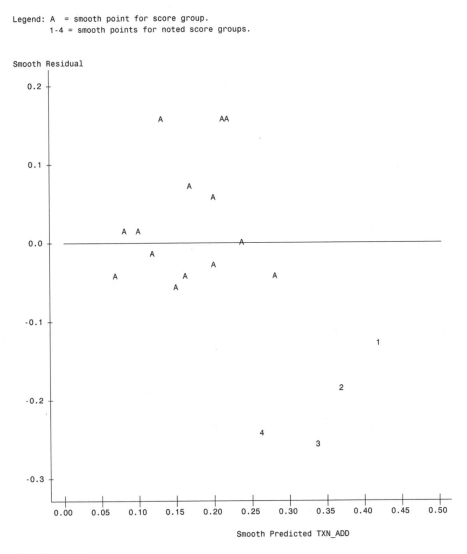

Legend: A = smooth point for score group.
        1-4 = smooth points for noted score groups.

FIGURE 3.11
Smooth Residual by Score Group Plot for Three Variable (FD2_RCP, MOS_OPEN and MOS_DUM) Model

to the null plot based General Association Test. Thus, the overall quality of prediction is considered good. That is, on average, the predicted TXN_ADD is equal to the actual TXN_ADD.

Easily seen, but not easily understood (at this point in the analysis), is the local pattern defined by four score groups (labeled 1 through 4) in the lower right-hand side of the plot. The local pattern explicitly shows that the smooth residuals are noticeably negative. The local pattern indicates a weak spot in the model, as its predictions for the individuals in the four score groups have, on average, a positive bias, that is, their predicted TXN_ADD tends to be larger than their actual TXN_ADD.

If implementation of the model can afford "exception rules" for individuals in a weak spot, then model performance can be enhanced. For example, response models typically have a weak spot, as prediction bias stems from limited information on new customers and outdated information on expired customers. Thus, if model implementation on a solicitation database can include exception rules, e.g., new customers are always targeted (assigned to the top decile), and expired customers are placed in the middle deciles, then the overall quality of prediction is improved.

For use in a later discussion, the descriptive statistics for the smooth residual by score groups/three-variable model plot are: for the smooth residuals, the minimum and maximum values and the range are –0.26 and 0.16, and 0.42, respectively; the standard deviation of the smooth residuals is 0.124.

3.13.2 Smooth Actual vs. Predicted by Decile Groups Plot

The *smooth actual vs. predicted by decile groups plot* is defined as the plot consisting of the mean actual response against the mean predicted response by *decile groups*. Decile groups are ten equal-sized classes, which are based on the predicted response values from the logistic regression model under consideration. Decile groupings is not an arbitrary partitioning of the data, as most database models are implemented at decile-level, and consequently built and validated at the decile-level.

The steps required to construct the smooth actual vs. predicted by decile groups plot and its interpretation are as follows:

1. *Score* the data by appending the predicted logit as outlined in Section 3.1.2.

2. *Convert* the predicted logit to predicted probability of response as outlined in Section 3.1.2.

3. *Determine* the decile groups. Rank in descending order the scored data by the predicted response values. Then, divide the scored-ranked data into ten equal-sized classes. The first class has the largest mean predicted response, labeled "top;" the next class is labeled "2," and so on. The last class has the smallest mean predicted response, labeled for "bottom."

4. For each decile group, *calculate* the mean (smooth) actual response and mean (smooth) predicted response, producing a set of ten smooth points, (smooth actual response, smooth predicted response).

5. *Plot* the smooth points by decile group, labeling the points by decile group.

6. *Draw* the 45-degree line on the plot. This line serves as a reference for assessing the quality of predictions at the decile-group level. If the smooth points are either on or *hug the 45-degree line* in their proper order (top-to-bot, or bot-to-top), then predictions, on average, are considered *good*.

7. *Determine* the "tightness" of the hug of the smooth points about the 45-degree line. To aid in an objective examination of the smooth plot, use the correlation coefficient between the smooth actual and predicted response points. The correlation coefficient serves as an indicator of the amount of scatter-about the 45-degree straight line. The larger the correlation coefficient, the lesser the scatter, the better the overall quality of prediction.

8. As previously discussed in Section 3.5.3, the correlation coefficient based on "big" points tends to produce a "big" r value, which serves as a gross estimate of the individual-level r value. The correlation coefficient based on smooth actual and predicted response points is a gross measure of the model's individual-level predictions. It is best served as a *comparative indicator* in choosing the better model.

3.13.2.1 Smooth Actual vs. Predicted by Decile Groups Plot for Case Study

I construct the smooth actual vs. predicted by decile groups plot based on Table 3.15 to determine the quality of the three-variable model predictions. The smooth plot in Figure 3.12 has a minimal scatter of the ten smooth points about the 45-degree lines, with two noted exceptions. Decile groups 4 and 6 appear to be the farthest away from the line (in terms of perpendicular

TABLE 3.15

Smooth Points by Deciles from Model Based on FD_RCP, MOS_OPEN , and MOS_DUM

		TXN_ADD		Predicted TXN_ADD	
DECILE	N	MEAN	MEAN	MIN	MAX
Top	492	0.069	0.061	0.061	0.061
2	493	0.047	0.061	0.061	0.061
3	493	0.037	0.061	0.061	0.061
4	492	0.089	0.080	0.061	0.085
5	493	0.116	0.094	0.085	0.104
6	493	0.085	0.104	0.104	0.104
7	492	0.142	0.118	0.104	0.121
8	493	0.156	0.156	0.121	0.196
9	493	0.185	0.198	0.196	0.209
Bottom	492	0.270	0.263	0.209	0.418
Total	4,926	0.119	0.119	0.061	0.418

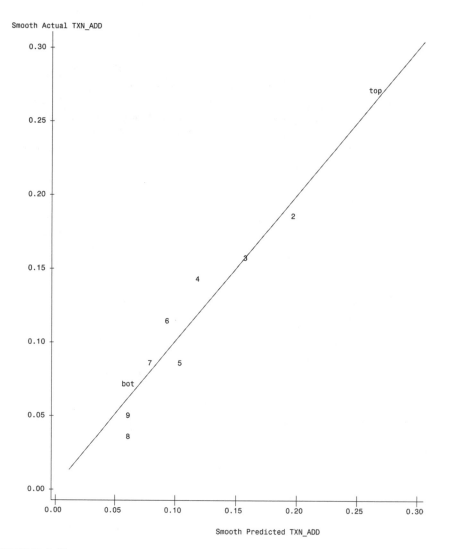

FIGURE 3.12

Smooth Actual vs. Predicted by Decile Group Plot for Three Variable (FD2_RCP, MOS_OPEN and MOS_DUM) Model

distance). Decile groups 8, 9 and "bot" are on top of each other, which indicates that the predictions are the same for these groups. The indication is that the model cannot discriminate among the least responding individuals. But, because implementation of response models typically exclude the lower three or four decile groups, their spread about 45-degree line and their (lack of) order are not as critical a feature in assessing the quality of the prediction. Thus, the overall quality of prediction is considered good.

The descriptive statistic for the smooth actual vs. predicted by decile groups/three-variable model plot is: the correlation coefficient between the smooth points, $r_{\text{sm. actual, sm. predicted: decile group}}$, is 0.972.

3.13.3 Smooth Actual vs. Predicted by Score Groups Plot

The *smooth actual vs. predicted by score groups plot* is defined as the plot consisting of the mean actual against the means predicted response by the score groups. Its construction and interpretation are virtually identical to the smooth actual vs. predicted by decile groups plot. The painlessly obvious difference is that decile groups are replaced by score groups, which are defined in the discussion in Section 3.13.1 on the smooth residual by score groups plot.

I compactly outline the steps for the construction and interpretation of the smooth actual vs. predicted by score groups plot:

1. *Score* the data by appending the predicted logit, and convert the predicted logit to predicted probability of response.
2. *Determine* the score groups, and calculate their smooth values for actual response and predicted response.
3. *Plot* the smooth actual and predicted points by score group.
4. *Draw* 45-degree line on the plot. If the smooth plot looks like the null plot, then it is concluded that the model predictions aggregated at the score group-level are considered good.
5. *Use* the correlation coefficient between the smooth points to aid in an objective examination of the smooth plot. This serves as an indicator of the amount of scatter-about the 45-degree line. The larger the correlation coefficient, the lesser the scatter, the better the overall quality of predictions. The correlation coefficient is best served as a comparative measure in choosing the better model.

3.13.3.1 Smooth Actual vs. Predicted by Score Groups Plot for Case Study

I construct the smooth actual vs. predicted by score groups plot based on Table 3.16 to determine the quality of the three-variable model predictions. The smooth plot in Figure 3.13 indicates that the scatter of the eighteen smooth points about the 45-degree line is good, except for the four points on the right-hand side of the line, labeled numbers 1 through 4. These points correspond to the four score groups, which became noticeable in the smooth residual plot in Figure 3.11. The indication is the same as that of the smooth residual plot: the overall quality of the prediction is considered good. However, if implementation of the model can afford exception rules for individuals who look like the four score groups, then the model performance can be improved.

TABLE 3.16

Smooth Points by Score Groups from Model Based on FD2_RCP, MOS_OPEN and MOS_DUM

			TXN_ADD	PROB_HAT
MOS_OPEN	FD2_OPEN	N	MEAN	MEAN
1	1	161	0.267	0.209
	2	56	0.268	0.359
	3	20	0.350	0.418
2	1	186	0.145	0.157
	2	60	0.267	0.282
	3	28	0.214	0.336
3	1	211	0.114	0.116
	2	62	0.274	0.217
	3	19	0.158	0.262
4	1	635	0.087	0.085
	2	141	0.191	0.163
	3	50	0.220	0.200
5	1	1,584	0.052	0.061
	2	293	0.167	0.121
	3	102	0.127	0.150
6	1	769	0.109	0.104
	2	393	0.186	0.196
	3	156	0.237	0.238
Total		4,926	0.119	0.119

The profiling of the individuals in the score groups is immediate from Table 3.16. The original predictor variables, instead of the re-expressed versions, are used to make the interpretation of the profile easier. The sizes — 20, 56, 28, 19 — of the four noticeable groups are quite small, for groups 1 to 4, respectively, which may account for the undesirable spread about the 45-degree line. However, there are three other groups of small size, 60, 62 and 50, which do not have noticeable spread about the 45-degree line. So, perhaps group size is not the reason for the undesirable spread. Regardless of why the unwanted spread exists, the four noticeable groups indicate that the three-variable model reflects a small weak spot, a segment that accounts for only 2.5% (= (20+56+28+19)/4926)) of the database population from which the sample was drawn. Thus, implementation of the three-variable model is expected to yield good predictions, even if exception rules cannot be afforded to the weak-spot segment, as its effects on model performance should hardly be noticed.

The descriptive profile of the weak-spot segment is as follows: newly opened accounts (less than 6 months) of customers with two or three accounts; recently opened accounts (between 6 months to 1 year) of customers with three accounts; and older accounts (between 1 to 1½ years) of customers with three accounts. The actual profile cells are:

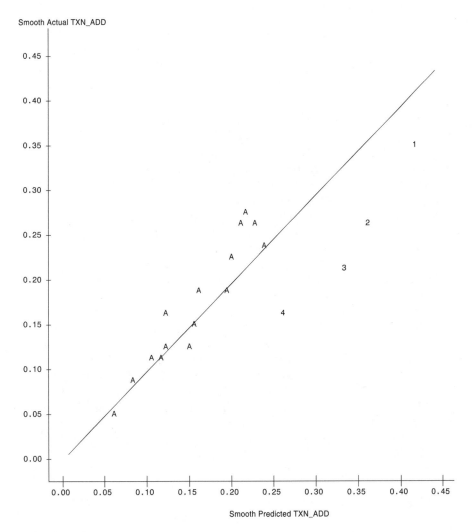

Legend: A = smooth point for score group.
        1-4 = smooth points for noted score groups.

FIGURE 3.13
Smooth Actual vs. Predicted by Score Group Plot for Three Variable (FD2_RCP, MOS_OPEN and MOS_DUM) Model

1. MOS_OPEN = 1 and FD2_OPEN = 3
2. MOS_OPEN = 1 and FD2_OPEN = 2
3. MOS_OPEN = 2 and FD2_OPEN = 3
4. MOS_OPEN = 3 and FD2_OPEN = 3

The descriptive statistic for the smooth actual vs. predicted by score groups/ three-variable model plot is: the correlation coefficient between the smooth points, r $_{sm. actual, sm. predicted: score group}$, is 0.848.

3.14 Evaluating the Data Mining Work

To appreciate the data mining analysis, which produced the three-variable EDA-model, I build a nonEDA model for comparison. I use the ever-popular stepwise logistic regression variable selection process, which the experienced analyst knows has a serious weakness. The stepwise and other statistics-based variable selection procedures can be classified as minimal data mining techniques, as they only find the "best" subset among the original variables without generating potentially important variables. They do *not* generate structure in the search for the best subset of variables. Specifically, they do not create new variables like re-expressed versions of the original variables, or derivative variables like dummy variables defined by the original variables. In contrast, the most productive data mining techniques generate structure from the original variables, and determine the best combination of those structures along with the original variables. More about variable selection in Chapter 16.

I perform a stepwise logistic regression analysis on TXN_ADD with the original five variables. The analysis identifies the best nonEDA subset consisting of only two variables: FD2_OPEN and MOS_OPEN; the output is Table 3.17. The G/df value is 61.3 (= 122.631), which is comparable to the G/df value (62.02) of the three-variable (FD2_RCP, MOS_OPEN, MOS_DUM) EDA model. Based on the G/df indicator of Section 3.9.4, I cannot declare that the three-variable EDA model is better than the two-variable nonEDA model.

TABLE 3.17

Best NonEDA Model Criteria for Assessing Model Fit

	Intercept Only	Intercept and All Variables	All Variables	
–2LL	3606.488	3483.857	122.631 with 2 df (p=0.0001)	
Variable	*Parameter Estimate*	*Standard Error*	*Wald Chi-Square*	*Pr > Chi-Square*
INTERCEPT	–2.0825	0.1634	162.3490	0.0001
FD2_OPEN	0.6162	0.0615	100.5229	0.0001
MOS_OPEN	–0.1790	0.0299	35.8033	0.0001

Could it be that all the EDA detective work was for naught — that the "quick and dirty" nonEDA model was the obvious one to build? The answer is no. Remember that an indicator is sometimes just an indicator that serves as a pointer to the next thing, such as moving on to the ladder of powers. Sometimes it is an instrument for automatically making a decision based on visual impulse, such as determining if the relationship is straight enough, or the scatter in a smooth residual plot is random. And sometimes, a lowly indicator does not have the force of its own to send a message until it is in the company of other indicators, e.g., smooth plots and their aggregate-level correlation coefficients.

I perform a simple comparative analysis of the descriptive statistics stemming from the EDA and nonEDA models to determine the better model. I need only to construct the three smooth plots — smooth residuals at the score-group level, smooth actuals at the decile-group level, and smooth actuals at the score-group level — from which I can obtain the descriptive statistics for the latter model, as I already have the descriptive statistics for the former model.

3.14.1 Comparison of Smooth Residual by Score Groups Plots: EDA vs. NonEDA Models

I construct the smooth residual by score groups plot in Figure 3.14 for the nonEDA model. The plot is not equivalent to the null plot based on the General Association Test. Thus, the overall quality of the nonEDA model's predictions is not considered good. There is a local pattern of five smooth points in the lower right-hand corner below the zero-line. The five smooth points, labeled I, II, III, IV and V, indicate that the predictions for the individuals in the five score groups have, on average, a positive bias; that is, their predicted TXN_ADD tends to be larger than their actual TXN_ADD. There is a smooth point, labeled VI, at the top of the plot that indicates a group of the individuals with an average negative bias. That is, their predicted TXN_ADD tends to be smaller than their actual TXN_ADD.

The descriptive statistics for the smooth residual by score groups/nonEDA model plot are as follows: for the smooth residual, the minimum and maximum values and the range are -0.33 and 0.29, and 0.62. The standard deviation of the smooth residuals is 0.167.

The comparison of the EDA and nonEDA smooth residuals indicates that the EDA model produces smaller smooth residuals (errors). The EDA smooth residual range is noticeably smaller than that of the nonEDA: 32.3% (= $(0.62 - 0.42)/0.62$) smaller. The EDA smooth residual standard deviation is noticeably smaller than that of the nonEDA: 25.7% (= $(0.167 - 0.124)/0.167$) smaller. The implication is that the EDA model has a better quality of prediction.

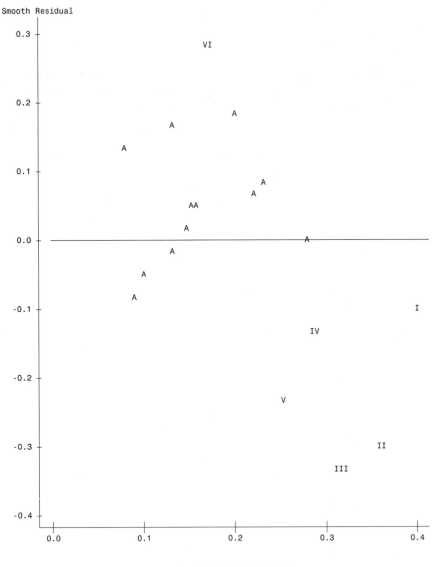

FIGURE 3.14
Smooth Residual by Score Group Plot for NonEDA (FD2_OPEN, MOS_OPEN) Model

3.14.2 Comparison of Smooth Actual vs. Predicted by Decile Groups Plots: EDA vs. NonEDA Models

I construct the smooth actual vs. predicted by decile groups plot in Figure 3.15 for the nonEDA model. The plot clearly indicates a scatter that does not hug the 45-degree line well, as decile groups "top" and 2 are far from the line, and 8, 9 and "bot" are out of order, especially "bot." The decile-based correlation coefficient between smooth points, $r_{sm.\ actual,\ sm.\ predicted:\ decile\ group}$, is 0.759.

The comparison of the decile-based correlation coefficients of the EDA and nonEDA indicates that the EDA model produces larger tighter hug about the 45-degree line. The EDA correlation coefficient is noticeably larger than that of the nonEDA: 28.1% (= (0.972 – 0.759)/0.759) larger. The implication is that EDA model has a better quality of prediction at the decile-level.

3.14.3 Comparison of Smooth Actual vs. Predicted by Score Groups Plots: EDA vs. NonEDA Models

I construct the smooth actual vs. predicted by scores groups plot in Figure 3.16 for the nonEDA model. The plot clearly indicates a scatter that does not hug the 45-degree line well, as scores groups II, III, V and VI are far from the line. The score group-based correlation coefficient between smooth points, $r_{sm.\ actual,\ sm.\ predicted:\ score\ group}$, is 0.635.

The comparison of the score group-based correlation coefficient of the EDA and nonEDA smooth actual plots indicate that the EDA model produces a tighter hug about the 45-degree line. The EDA correlation coefficient is noticeably larger than that of the nonEDA: 33.5% (= (0.848 – 0.635)/0.635) larger. The implication is that the EDA model has a better quality of prediction at the score-group level.

3.14.4 Summary of the Data Mining Work

From the comparative analysis, I have the following:

1. The overall quality of the EDA model's predictions is better than that of the nonEDA, as the former model's smooth residual plot is null, and that of the nonEDA model is not.

2. The EDA model's prediction errors are smaller than those of the nonEDA, as the former model's smooth residuals have less spread (smaller range and standard deviation). In addition, the EDA model has better aggregate-level predictions than that of the nonEDA, as the former model has less prediction bias (larger correlations between smooth actual and predicted values at decile- and score-group levels).

Legend: Values top, 2, ⋯, 9, bot are decile groups.

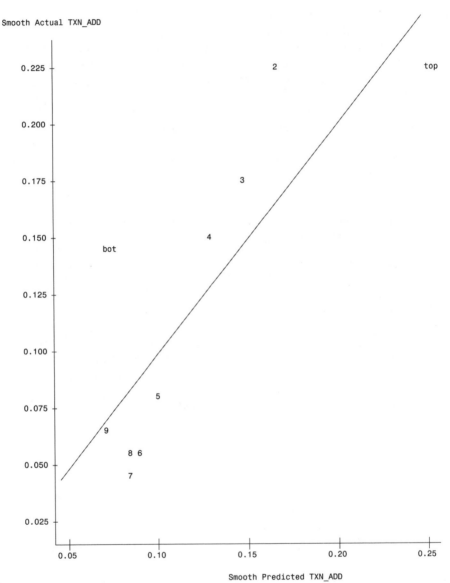

FIGURE 3.15
Smooth Actual vs. Predicted by Decile Group Plot for NonEDA (FD2_OPEN, MOS_OPEN)
Model

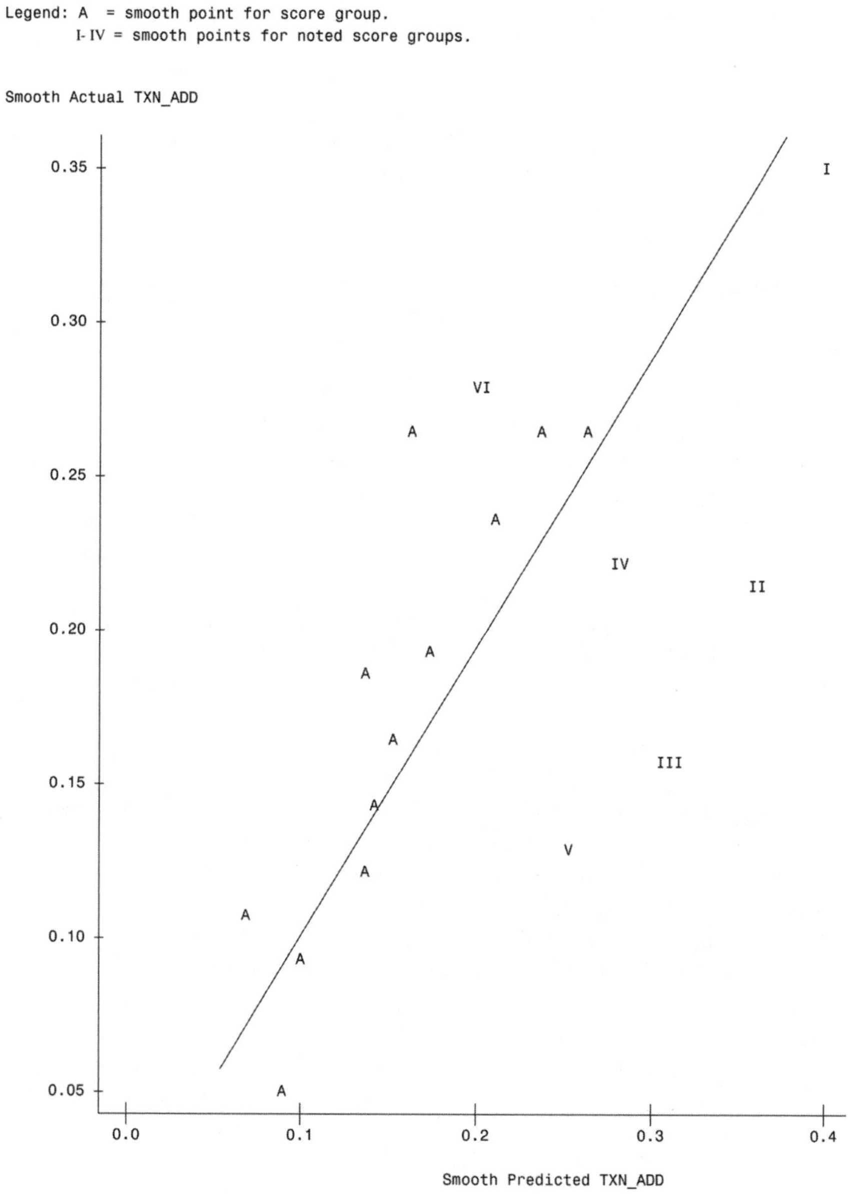

Legend: A = smooth point for score group.
        I- IV = smooth points for noted score groups.

Smooth Actual TXN_ADD

Smooth Predicted TXN_ADD

FIGURE 3.16
Smooth Actual vs. Predicted by Score Group Plot for NonEDA (FD2_OPEN, MOS_OPEN Model

3. I conclude the three-variable EDA model, consisting of FD2_RCP, MOS_OPEN and MOS_DUM is better than the two-variable non-EDA model, consisting of FD2_OPEN and MOS_OPEN.

As the last effort to improve the EDA model, I consider the last unmined candidate predictor variable FD_TYPE in the next section.

3.15 Smoothing a Categorical Variable

The classic approach to include a categorical variable into the modeling process involves *dummy variable coding*. A categorical variable with k classes of qualitative (nonnumerical) information is replaced by a set of k-1 quantitative dummy variables. The dummy variables are defined by the present or absent of the class values. The class left out is called the reference class, to which the other classes are compared when interpreting the effects of dummy variables on response. The classic approach instructs that the complete set of k-1 dummy variables is included in the model regardless of the number of dummy variables that are declared nonsignificant. This approach is problematic when the number of classes is large, which is typically the case in big data applications. By chance alone, as the number of classes increases, the probability of one or more dummy variables being declared nonsignificant increases. To put all the dummy variables in the model effectively adds "noise" or unreliability to the model, as nonsignificant variables are known to be "noisy." Intuitively, a large set of inseparable dummy variables poses a difficulty in model building, in that they quickly "fill up" the model not allowing room for other variables.

The EDA approach of treating a categorical variable for model inclusion is a viable alternative to the classic approach, as it explicitly addresses the problems associated with a large set of dummy variables. It reduces the number of classes by merging (smoothing or averaging) the classes with comparable values of the target variable under study, which for the application of response modeling is the response rate. The *smoothed* categorical variable, now with fewer classes, is less likely to add noise in the model and allows more room for other variables to get into the model.

There is an additional benefit offered by smoothing of a categorical variable. The information captured by the smoothed categorical variable tends to be more reliable than that of the complete set of dummy variables. The reliability of information of the categorical variable is only as good as the aggregate reliability of information of the individual classes. Classes of small size tend to provide unreliable information. Consider the extreme situation of a class of size one. The estimated response rate for this class is either 100% or 0% because the sole individual either responds or does not respond, respectively. It is unlikely that the estimated response rate is the true response rate for this class. This class is considered to provide unreliable information as to its true response rate. Thus, the reliability of information for the categorical variable itself decreases as the number of small classes increases. The smoothed categorical variable tends to have greater reliability than the set of dummy variables because it intrinsically has fewer classes, and, consequently, has larger class sizes due to the merging process. EDA's rule-of-thumb for small class size: less than 200 is considered small.

CHAID is often the preferred EDA technique for smoothing a categorical variable. In essence, CHAID is an excellent EDA technique, as it involves the three main elements of statistical detective work, "numerical, counting and graphical." CHAID forms new larger classes based on a numerical merging, or averaging of response rates, and counts the reduction in the number of classes as it determines the best set of merged classes. Lastly, CHAID's output is conveniently presented in an easy to read and understand graphical display, a treelike box diagram with leaf-boxes representing the merged classes.

The technical details of CHAID's merging process is beyond the scope of this chapter. CHAID will be covered in detail in subsequent chapters, so here I will briefly discuss and illustrate it with the smoothing of the last variable to be considered for predicting TXN_ADD response, namely, FD_TYPE.

3.15.1 Smoothing FD_TYPE with CHAID

Remember that FD_TYPE is a categorical variable that represents the product type of the customer's most recent investment purchase. It assumes fourteen products (classes) coded A, B, C, ..., N. The TXN_ADD response rate by FD_TYPE values are in Table 3.18.

There are seven small classes, F, G, J, K, L, M and N, with sizes, 42, 45, 57, 94, 126, 19 and 131, respectively. Their response rates — 0.26, 0.24, 0.19, 0.20, 0.22, 0.42 and 0.16, respectively — can be considered potentially unreliable. Class B has the largest size, 2828, with a surely reliable 0.06 response rate. The remaining six presumably reliable classes, A, C, D, E, H have sizes between 219 to 368.

The CHAID tree for FD_TYPE in Figure 3.17 is read and interpreted as follows:

1. The top box, the root of the tree, represents the sample of 4,926 with response rate 11.9%.

2. The CHAID technique smoothes FD_TYPE by way of merging the original fourteen classes into three merged (smoothed) classes, as displayed in the CHAID tree with three leaf-boxes.

3. The left-most leaf, which consists of the seven small unreliable classes and the two reliable classes A and E, represents a newly merged class with a reliable response rate 24.7% based on a class size of 1,018. In

TABLE 3.18

FD_TYPE

FD_TYPE	TXN_ADD	
	N	MEAN
A	267	0.251
B	2,828	0.066
C	250	0.156
D	219	0.128
E	368	0.261
F	42	0.262
G	45	0.244
H	225	0.138
I	255	0.122
J	57	0.193
K	94	0.202
L	126	0.222
M	19	0.421
N	131	0.160
Total	4,926	0.119

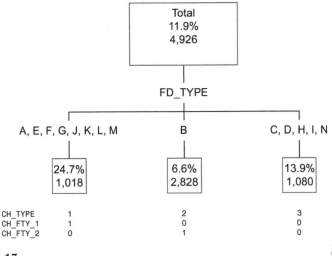

FIGURE 3.17
Double Smoothing of FD_TYPE with CHAID

this situation, the smoothing process increases the reliability of the small classes with a two-step averaging. The first step combines all the small classes into a temporary class, which by itself produces a reliable average response rate of 22.7% based on a class size of 383. In the second step, which does not always occur in smoothing, the temporary class is further united with the already reliable classes A and E, because the latter classes have comparable response rates to the temporary class response rate. The *double-smoothed* newly merged class represents the average response rate of the seven small classes and classes A and E. In situations where double-smoothing does not occur, the temporary class is the final class.

4. The increased reliability that smoothing of a categorical variable offers can now be clearly illustrated. Consider class M with its unreliable estimated response rate 42% based on class size 19. The smoothing process puts class M in the larger, more reliable left-most leaf with response rate 24.7%. The implication is that class M now has a more reliable estimate of response rate, namely, the response rate of its newly assigned class, 24.7%. Thus, the smoothing has effectively adjusted class M's original estimated response rate downward, from a positively biased 42% to a reliable 24.7%. In contrast, within the same smoothing process, class J's adjustment is upward, from a negatively biased 19% to 24.7%. Not surprising, the two reliable classes A and E remain noticeably unchanged, from 25% and 26% to 24.7%, respectively.

5. The middle leaf consists of only class B, defined by a large class size of 2,828 with a reliable response rate of 6.6%. Apparently, class B's low response rate is not comparable to any class (original, temporary or newly merged) response rate to warrant a merging. Thus, class B's original estimated response rate is unchanged after the smoothing process. This presents no concern over the reliability of class B because its class size is largest from the outset.

6. The right-most leaf consists of large classes C, D, H and I, and the small class N for an average reliable response rate of 13.9% with class size 1,080. The smoothing process adjusts class N's response rate downward, from 16% to a smooth 13.9%. The same adjustment occurs for class C. The remaining classes D, H and I, experience an upward adjustment.

I call the smoothed categorical variable CH_TYPE. Its three classes are labeled 1, 2 and 3, corresponding to the leaves from left to right, respectively (see bottom of Figure 3.7). I also create two dummy variables for CH_TYPE:

1. CH_FTY_1 = 1 if FD_TYPE = A, E, F, G, J, K, L or M; otherwise, CH_FTY_1 = 0;

2. CH_FTY_2 = 1 if FD_TYPE = B; otherwise, CH_FTY_2 = 0.

3. This dummy variable construction uses class CH_TYPE = 3 as the reference class. If an individual has values CH_FTY_1 = 0 and CH_FTY_2 = 0, then the individual implicitly has CH_TYPE = 3 and has one of the original classes, C, D, H, I or N.

TABLE 3.19

G and df for CHAID-smoothed FD_TYPE

Variable	−2LL	G	df	p-value
INTERCEPT	3606.488			
CH_FTY_1 & CH_FTY_2	3390.021	216.468	2	0.0001

3.15.2 Importance of CH_FTY_1 and CH_FTY_2

I assess the importance of the CHAID-based smoothed variable CH_TYPE by performing a logistic regression analysis on TXN_ADD with both CH_FTY_1 and CH_FTY_2, as the set dummy variable must be together in the model; the output is in Table 3.19. The G/df value is 108.234 (= 216.468/ 2), which is greater than the standard G/df value 4. Thus, CHFTY_1 and CH_FTY_2 together are declared important predictor variables of TXN_ADD.

3.16 Additional Data Mining Work for Case Study

I try to improve the predictions of the three-variable (MOS_OPEN, MOS_DUM and FD2_RCP) model with the inclusion of the smoothed variable CH_TYPE. I perform the logistic regression model on TXN_ADD with MOS_OPEN, MOS_DUM, FD2_RCP, and CH_FTY_1 and CH_FTY_2; the output is in Table 3.20. The Wald Chi-square value for FD2_RCP is less the 4. Thus, I delete FD2_RCP from the model, and rerun the model with the remaining four variables.

TABLE 3.20

Logistic Model: EDA Model Variables plus CH_TYPE Variables

	Intercept Only	Intercept and All Variables	All Variables	
−2LL	3606.488	3347.932	258.556 with 5 df (p = 0.0001)	
Variable	*Parameter Estimate*	*Standard Error*	*Wald Chi-Square*	*Pr > Chi-Square*
INTERCEPT	−0.7497	0.2464	9.253	0.0024
CH_FTY_1	0.6264	0.1175	28.4238	0.0001
CH_FTY_2	−0.6104	0.1376	19.6737	0.0001
FD2_RCP	0.2377	0.2212	1.1546	0.2826
MOS_OPEN	−0.2581	0.0398	42.0054	0.0001
MOS_DUM	0.7051	0.1365	26.6804	0.0001

TABLE 3.21

Logistic Model: Four Variable EDA Model

	Intercept Only	Intercept and All Variables	All Variables	
−2LL	3606.488	3349.094	257.395 with 4 df (p = 0.0001)	
Variable	Parameter Estimate	Standard Error	Wald Chi-Square	Pr > Chi-Square
INTERCEPT	−0.9446	0.1679	31.6436	0.0001
CH_FTY_1	0.6518	0.1152	32.0362	0.0001
CH_FTY_2	−0.6843	0.1185	33.3517	0.0001
MOS_OPEN	−0.2510	0.0393	40.8141	0.0001
MOS_DUM	0.7005	0.1364	26.3592	0.0001

The four-variable (MOS_OPEN, MOS_DUM, CH_FTY_1 and CH_FTY_2) model produces comparable Wald Chi-square values for the four variables; the output is in Table 3.21. The G/df value equals 64.348 (= 257.395/4), which is slightly larger than the G/df (62.02) of the three-variable (MOS_OPEN, MOS_DUM, FD2_RCP) model. This is not a strong indication that the four-variable model has more predictive power than the three-variable model.

In the following four sections, I perform the comparative analysis, similar to the EDA vs. nonEDA analysis in Section 3.14, to determine whether the four variable (4var-) EDA model is better than the three variable (3var-) EDA model. I need the smooth plot descriptive statistics for the latter model, as I already have the descriptive statistics for the former model.

3.16.1 Comparison of Smooth Residual by Score Group Plots: 4var- vs. 3var-EDA Models

The smooth residual by score group plot for the 4var-EDA model in Figure 3.18 is equivalent to the null plot based on the General Association Test. Thus, the overall quality of the model's predictions is considered good. It is worthy of notice that there is a far-out smooth point, labeled FO, in the middle of the top of the plot. This smooth residual point corresponds to a score group consisting of 56 individuals (accounting for 1.1% of the data), indicating a weak spot.

The descriptive statistics for the smooth residual by score groups/4var-model plot are as follows: for the smooth residual, the minimum and maximum values and the range are −0.198 and 0.560, and 0.758, respectively; the standard deviation of the smooth residual is 0.163.

The descriptive statistics based on all the smooth points *excluding* the FO smooth point is worthy of notice because such statistics are known to be sensitive to far-out points, especially when they are smoothed and account for a very small percentage of the data. For the FO-adjusted smooth residual,

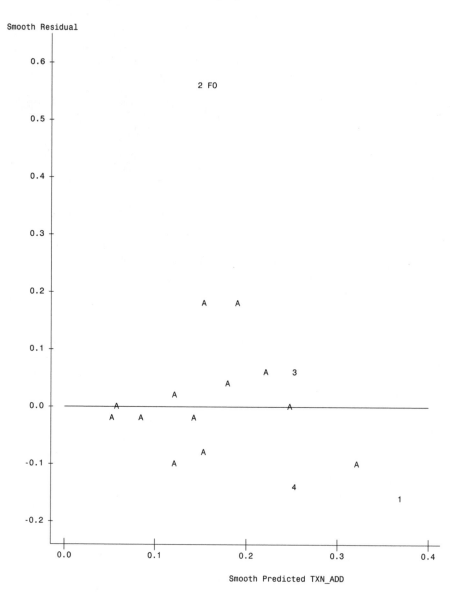

FIGURE 3.18
Smooth Residual by Score Group Plot for 4var-EDA (MOS_OPEN, MOS_DUM, CH_FTY_1 CH_FTY_2) Model

the minimum and maximum values and the range are –0.198 and 0.150, and 0.348, respectively; the standard deviation of the smooth residuals is 0.093.

The comparison of the 3var- and 4var-EDA model smooth residuals indicates that the former model produces smaller smooth residuals (errors). The 3var-EDA model smooth residual range is noticeably smaller than that of the latter model: 44.6% (= (0.758 – 0.42)/0.758) smaller. The 3var-EDA model smooth residual standard deviation is noticeably smaller than that of the 4var-EDA model: 23.9% (= (0.163 – 0.124/0.163) smaller. The implication is that the CHAID-based dummy variables carrying the information of FD_TYPE are not important enough to produce better predictions than that of the 3var-EDA model. In other words, the 3var-EDA model has a better quality of prediction.

However, if implementation of the TXN_ADD model permits an exception rule for the FO score group/weak spot, the implication is that the 4var-EDA model has a better quality of predictions, as the model produces smaller smooth residuals. The 4var-EDA model FO-adjusted smooth residual range is noticeably smaller than that of the 3var-EDA model: 17.1% (= (0.42 – 0.348)/0.42) smaller. The 4var-EDA model FO-adjusted smooth residual standard deviation is noticeably smaller than that of the 3var-EDA model: 25.0% (= (0.124 – 0.093)/0.124).

3.16.2 Comparison of Smooth Actual vs. Predicted by Decile Groups Plots: 4var- vs. 3var-EDA Models

The smooth actual vs. predicted by decile groups plot for the 4var-EDA model in Figure 3.19 indicates a very good hugging of scatter-about the 45-degree lines, despite the following two exceptions. First, there are two pairs of deciles groups, 6 and 7, and 8 and 9, where the decile groups in each pair are adjacent to each other. This indicates that the predictions are different for decile groups within each pair, which should have the same response rate. Second, the "bot" decile group is very close to the line, but out of order. Because implementation of responses model typically exclude the lower three or four decile groups, their spread about 45-degree line and their (lack of) order is not as critical a feature in assessing the quality of predictions. Thus, overall the plot is considered very good. The coefficient correlation between smooth actual and predicted points, $r_{\text{sm. actual, sm. predicted:decile group}}$, is 0.989.

The comparison of the decile-based correlation coefficient of the 3var- and 4var-EDA model smooth actual plots indicates that the latter model produces a meagerly tighter hug about the 45-degree line. The 4var-EDA model correlation coefficient is hardly noticeably larger than that of the three-variable model: 1.76% (= (0.989–0.972)/0.972) smaller. The implication is that both models have equivalent quality of prediction at the decile-level.

Legend: Values top, 2, ⋯, 9, bot are decile groups.

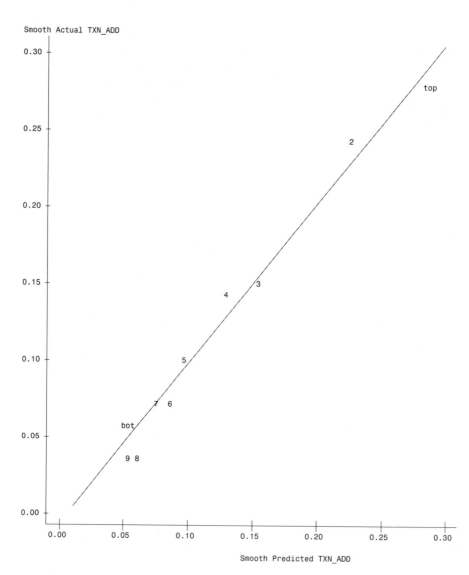

FIGURE 3.19

Smooth Actual vs. Predicted by Decile Group Plot for 4var-EDA (MOS_OPEN, MOS_DUM, CH_FTY_1 CH_FTY_2) Model

3.16.3 Comparison of Smooth Actual vs. Predicted by Score Groups Plots: 4var- vs. 3var-EDA Models

The score groups for the smooth actual vs. predicted by score groups plot for the 4var-EDA model in Figure 3.20 are defined by the variables in the 3var-

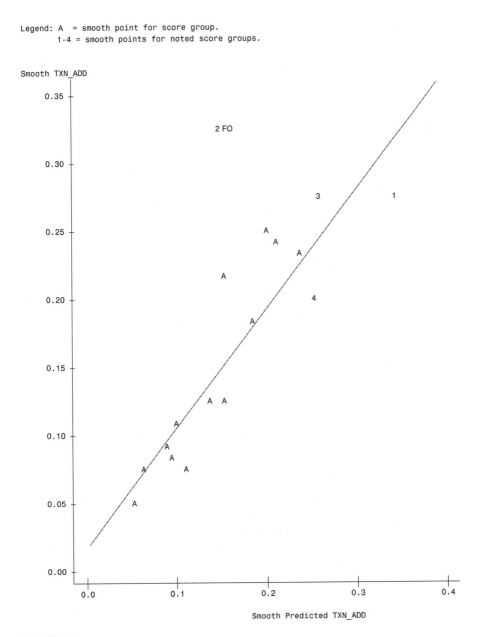

FIGURE 3.20
Smooth Actual vs. Predicted by Score Group Plot for 4var-EDA (MOS_OPEN, MOS_DUM, CH_FTY_1 CH_FTY_2) Model

EDA model to make an uncomplicated comparison. The plot indicates a very nice hugging about the 45-degree line, except for one far-out smooth point, labeled FO, which was initially uncovered by the smooth residual plot in Figure 3.18. The score group-based correlation coefficient between all smooth points, r sm. actual, sm. predicted:score group, is 0.784; the score group-based correlation without the far-out score group FO, r sm. actual, sm. predicted: score group-FO, is 0.915. The comparison of the score group-based correlation coefficient of the 3var- and 4var-smooth actual plots indicates that the 3var-EDA model produces a somewhat noticeably tighter hug about the 45-degree line. The 3var-EDA model score group-based correlation coefficient is somewhat noticeably larger that of the 4var-EDA model: 8.17% (= (0.848 – 0.784) –/0.784) larger. The implication is the 3var-EDA model has a somewhat better quality of prediction at the score group-level.

However, the comparison of the score group-based correlation coefficient of the 3var- and 4var-smooth actual plots *without the FO score group* produces a reverse implication. The 4var-EDA model score group-based correlation coefficient is somewhat noticeably larger than that of the three-variable model: 7.85% (= (0.915 0.848) –/0.848) larger. The implication is that the 4var-EDA model, without the FO-score group, has a somewhat better quality of prediction at the score group-level.

3.16.4 Final Summary of the Additional Data Mining Work

The comparative analysis offers the following:

1. The overall quality of the 3var- and 4var-EDA models are considered good, as both models have a null smooth residual plot. Worthy of notice, there is a very small weak spot (FO score group accounting for 1.1% of the data) in the latter model.

2. The 3var-EDA model's prediction errors are smaller than those of the 4var-EDA model, as the former model's smooth residuals have less spread (smaller range and standard deviation). The 3var-EDA model has equivalent/somewhat better aggregate-level predictions, as the former model has equivalent/somewhat-less prediction bias (equivalent/larger correlation between smooth actual and predicted values at the decile-level/score group-level).

3. If model implementation can accommodate an exception rule for the FO weak spot, the indicators suggest that the 4var-EDA model has less spread and somewhat-better aggregate-level predictions.

4. In sum, I prefer the 3var-EDA model, consisting of MOS_OPEN, MOS_DUM, and FD2_RCP. If exception rules for the far-out score group can be used and effectively developed, I prefer the four model, consisting of MOS_OPEN, MOS_DUM, CH_FTY_1 and CH_FTY_2.

3.17 Summary

The logistic regression model is presented as the workhorse of database response modeling. As such, I demonstrated how it fulfills the desired analysis of a Yes-No response variable, in that it provides individual probabilities of response, as well as yielding a meaningful aggregation of individual-level probabilities into decile-level probabilities of response. This is often required in database implementation of response models. Moreover, I showed the durability and usefulness of the 60-year-old analysis and modeling technique, as it works well within today's EDA/data mining paradigm.

I first illustrated the rudiments of the logistic regression model by discussing the SAS program code for building and scoring a logistic regression model. Working through a small data set, I pointed out — and ultimately clarified — an often vexing relationship between the actual and predicted response variables: the former assumes two nominal values, typically 1–0 for yes-no responses, respectively, yet the latter assumes "logits," which are continuous values between –7 and +7. This disquieting connection is currently nursed by the latest release of SAS (version 8). It is typical for the analyst to name the actual response variable "RESPONSE." In SAS logistic procedure, the naming convention of the predicted response variable is the name that the analyst assigns to the actual response variable, typically, "RESPONSE." Thus, the predicted response variable labeled "RESPONSE" may cause the analyst to think that the predicted response variable is a binary variable, not the logit it actually is.

Next, I presented a case study that serves as the vehicle for introducing a host of data mining techniques, which are mostly specific to the logistic regression model. The logistic regression model has the implied critical assumption that the underlying relationship between a given predictor variable and the logit of response is straight-line. I outline the steps for logit plotting to test the assumption. If the test results are negative, i.e., a straight line exists, then no further attention is required for that variable, which can now be included in the model. If the test result is positive, the variable needs to be straightened before it can justifiably be included in the model.

I discussed and illustrated two straightening methods — re-expressing with the Ladder of Powers and the Bulging Rule — both of which are applicable to all linear models like logistic regression and ordinary regression. The efficacy of these methods is determined by the well-known correlation coefficient, which is often misused. At this point, I reinstated the frequently overlooked assumption of the correlation coefficient; that is, that the underlying relationship at hand is a straight line. Additionally, I introduced an implied assumption of the correlation coefficient specific to its proper use, which is to acknowledge that the correlation coefficient for re-expressed variables with smoothed big data is a gross measure of straightness of re-expressed variables.

Continuing with the case study, I demonstrated a logistic regression-specific data mining alternative approach to the classical method of assessing. This alternative approach involves the importance of individual predictor variables, as well as the importance of a subset of predictor variables and the relative importance of individual predictor variables, in addition to the goodness of model predictions. Additional methods specific to logistic regression set out included: selecting the best subset of predictor variables, and comparing the importance between two subsets of predictor variables.

My illustration within the case study includes one last alternative method, applicable to all models, linear and nonlinear; that is, smoothing a categorical variable for model inclusion. The method re-expresses a set of dummy variables, which is traditionally used to include a categorical variable in a model, into a new parsimonious set of dummy variables that is more reliable than the original set, and easier to include in a model than the original set.

The case study produced a final comparison of a data-guided EDA model and a nonEDA model, based on the ever-popular stepwise logistic regression variable selection process, whose weaknesses were discussed. The EDA model was the preferred model with a better quality of prediction.

4

Ordinary Regression: The Workhorse of Database Profit Modeling

Ordinary regression is the most popular technique for predicting a quantitative outcome, such as profit or sales. It is considered the workhorse of *profit* modeling as its results are taken as the gold standard. Moreover, the ordinary regression model is used as the benchmark for assessing the superiority of new and improved techniques. In database marketing, profit (variously defined as any measure of an individual's valuable contribution to a business) due to a prior solicitation is the quantitative variable, and the ordinary regression model is built to predict an individual's profit from a future solicitation.

I provide a brief overview of ordinary regression and include the SAS program code for building and scoring an ordinary regression model. Then, I present a mini case study to illustrate that the data mining techniques presented in Chapter 3 carry over with minor modification to ordinary regression. Data analysts who are called upon to provide statistical support to managers monitoring expected revenue from marketing campaigns will find this chapter an excellent reference to profit modeling.

4.1 Ordinary Regression Model

Let Y be a quantitative dependent variable that assumes a continuum of values. The ordinary regression model, formally known as the ordinary least squares (OLS) regression model, predicts the Y value for an individual based on the values for predictor (independent) variables X_1, X_2, ..., X_n for that individual. The OLS model is defined in Equation (4.1):

$$Y = b_0 + b_1{}^*X_1 + b_2{}^*X_2 + ... + b_n{}^*X_n \qquad (4.1)$$

An individual's predicted Y value is calculated by "plugging-in" the values of the predictor variables for that individual in Equation (4.1). The bs are the OLS regression coefficients, which are determined by the calculus-based method of least squares estimation; the lead coefficient b_0 is referred to as the Intercept.

In practice the quantitative dependent variable does not have to assume a progression of values that vary by minute degrees. It can assume just several dozens of discrete values, and work quite well with the OLS methodology. When the dependent variable assumes only two values, the logistic regression model, not the ordinary regression model, is the appropriate technique. Even though logistic regression has been around for several decades, there is some misunderstanding over the practical (and theoretical) weakness of using the OLS model for a binary response dependent variable.

4.1.1   Illustration

Consider dataset A, which consists of ten individuals and three variables in Table 4.1: the quantitative variable PROFIT in dollars (Y), INCOME in thousand-dollars (X1), and AGE in years (X2). I regress PROFIT on INCOME and AGE using dataset A. The OLS output in Table 4.2 includes the ordinary regression coefficients and other "columns" of information. The Estimate column contains the coefficients for INCOME, AGE, and the INTERCEPT variables. The coefficient b_0 for the INTERCEPT variable is used as a "start" value given to all individuals, regardless of their specific values for the predictor variables in the model.

The estimated OLS PROFIT Model is defined in Equation (4.2):

$$PROFIT = 52.2778 + 0.2667*INCOME - 0.1622*AGE \qquad (4.2)$$

TABLE 4.1

Dataset A

Profit ($)	Income ($000)	Age (years)
78	96	22
74	86	33
66	64	55
65	60	47
64	98	48
62	27	27
61	62	23
53	54	48
52	38	24
51	26	42

TABLE 4.2

OLS Output: PROFIT with INCOME and AGE

Source	df	Sum of Squares	Mean Square	F Value	Pr > F
Model	2	460.3044	230.1522	6.01	0.0302
Error	7	268.0957	38.2994		
Corrected Total	9	728.4000			
		Root MSE	6.18865	R-Square	0.6319
		Dependent Mean	62.60000	Adj R-Sq	0.5268
		Coeff Var	9.88602		

Variable	df	Parameter Estimate	Standard Error	t Value	Pr > |t|
INTERCEPT	1	52.2778	7.7812	7.78	0.0003
INCOME	1	0.2669	0.2669	0.08	0.0117
AGE	1	–0.1622	–0.1622	0.17	0.3610

4.1.2 Scoring A OLS Profit Model

The SAS-code program in Figure 4.1 produces the OLS Profit Model built with dataset A, and scores the external dataset B in Table 4.3. The SAS procedure REG produces the ordinary regression coefficients and puts them in the "ols_coeff" file, as indicated by the code "outest = ols_coeff." The ols_coeff file produced by SAS (versions 6 and 8) is in Table 4.4.

The SAS procedure SCORE scores the five individuals in dataset B using the OLS coefficients, as indicated by the code "score = ols_coeff." The procedure appends the predicted PROFIT variable in Table 4.3 (called pred_PROFIT as indicated by "pred_PROFIT" in the second line of code in Figure 4.1) to the output file B_scored, as indicated by the code "out = B_scored."

```
/****** Building the OLS Profit Model on dataset A ************/
PROC REG  data = A  outest = ols_coeff;
pred_Profit: model Profit =
Income Age;
run;

/****** Scoring the OLS Profit Model on dataset B ************/
PROC SCORE data = B predict type = parms  score = ols_coeff
out = B_scored;
var Income Age;
run;
```

FIGURE 4.1
SAS Code for Building and Score OLS Profit Model

TABLE 4.3

Dataset B

Income ($000)	Age (years)	Predicted Profit ($)
148	37	85.78
141	43	82.93
97	70	66.81
90	62	66.24
49	42	58.54

TABLE 4.4

OLS_Coeff File (SAS ver.6 and 8)

OBS	_MODEL_	_TYPE_	_DEPVAR_	_RMSE_	Intercept	Income	Age	Profit
1	est_Profit	PARMS	Profit	6.18865	52.52778	0.26688	-0.16217	-1

4.2   Mini Case Study

I present a "big" discussion on ordinary regression modeling with the mini dataset A. I use this extremely small data to not only make the discussion of the data mining techniques tractable, but to emphasize two aspects of data mining. First, data mining techniques of great service should work as well with small data as with big data, as explicitly stated in the definition of data mining in Chapter 1. Second, every fruitful effort of data mining on small data is evidence that big data are not always necessary to uncover structure in the data. This evidence is in keeping with the EDA philosophy that the data analyst should work from simplicity until indicators emerge to go further: if predictions are not acceptable, then increase data size.

The objective of the mini case study is as follows: to build a PROFIT Model based on INCOME and AGE. The ordinary regression model (celebrating almost two hundred years of popularity since the invention of the method of least squares on March 6, 1805) is the quintessential linear model, which implies the all-important assumption that the underlying relationship between a given predictor variable and the dependent variable is linear. Thus, I use the method of smoothed scatterplots, as described in Chapter 2, to determine whether the linear assumption holds for PROFIT with INCOME and with AGE. For the mini dataset, the smoothed scatterplot is defined by ten slices, each of size 1. Effectively, the smooth scatterplot is the simple scatterplot of ten paired (PROFIT, Predictor Variable) points. (Contrasting note: the logit plot as discussed with the logistic regression in Chapter 3 is neither possible nor relevant with OLS methodology. The quantitative dependent variable does not require a transformation, like converting logits into probabilities as found in logistic regression.)

4.2.1   Straight Data for Mini Case Study

Before proceeding with the analysis of the mini case study, I clarify the use of the Bulging Rule when analysis involves OLS regression. The Bulging Rule states that the data analyst should try re-expressing the predictor variables as well as the dependent variable. As discussed in Chapter 3, it is not possible to re-express the dependent variable in a logistic regression analysis. However, in performing an ordinary regression analysis, re-expressing the dependent variable is possible, but the Bulging Rule needs to be supplemented. Consider the illustration discussed below.

A data analyst is building a profit model with the quantitative dependent variable Y, and three predictor variables X_1, X_2 and X_3. Based on the Bulging Rule, the analyst determines that the powers of $\frac{1}{2}$ and 2 for Y and X_1, respectively, produce a very adequate straightening of the Y-X_1 relationship. Let's assume that the correlation between the square root Y (sqrt_Y) and square of X_1 (sq_X_1) has a reliable r $_{sqrt\_Y,\ sq\_X1}$ value of 0.85.

Continuing this scenario, she determines that the powers of 0 and $\frac{1}{2}$, and $-\frac{1}{2}$ and 1, for Y and X_2, and Y and X_3, respectively, also produce very adequate straightening of the Y-X_2 and Y-X_3 relationships, respectively. Let's assume the correlations between the log of Y (log_Y) and the square root of X_2 (sq_X_2), and between the negative square root of Y (negsqrt_Y) and X_3 have reliable $r_{\text{log\_Y, sq\_X1}}$ and $r_{\text{negsqrt\_Y, X3}}$ values of 0.76 and 0.69, respectively. In sum, the data analyst has the following results:

1. Best relationship between square root Y (p = $\frac{1}{2}$) and square of X_1 has $r_{\text{sqrt\_Y, sq\_X1}} = 0.85$.
2. Best relationship between log of Y (p = 0) and square root of X_2 has $r_{\text{log\_Y, sq\_X2}} = 0.76$.
3. Best relationship between negative square root of Y (p = $-\frac{1}{2}$) and X_3 has $r_{\text{neg\_sqrt\_Y, X3}} = 0.69$.

In pursuit of a good OLS profit model, the following guidelines have proven valuable when several re-expressions of the quantitative dependent variable are suggested by the Bulging Rule.

1. If there is a small range of *dependent-variable powers* (powers used in re-expressing the dependent variable), then the best re-expressed dependent variable is the one with the noticeably largest correlation coefficient. In the illustration, the best re-expression of Y is the square root Y, as its correlation has the largest value, $r_{\text{sqrt\_Y, sq\_X1}}$ equals 0.85. Thus, the data analyst builds the model with the square root Y and the square of X_1, and needs to re-express X_2 and X_3 again with respect to the square root of Y.

2. If there is a small range of dependent-variable powers and the correlation coefficient values are comparable, then the best re-expressed dependent variable is defined by the *average power* among the dependent variable powers. In the illustration, if the data analyst were to consider the r values (0.85, 0.76 and 0.69) comparable, then the average power would be 0, which is one of the powers used. Thus, the data analyst builds the model with the log of Y and square root of X_2 and needs to re-express X_1 and X_3 again with respect to the log of Y.

   If the average power is not one of the dependent-variable powers used, then all predictor variables would need to be re-expressed again with the newly assigned re-expressed dependent variable, Y raised to the "average power."

3. When there is a large range of dependent-variable powers, which is likely when there are many predictor variables, the practical and productive approach to the Bulging Rule when building an OLS profit model consists of initially re-expressing only the predictor

variables, leaving the dependent variable unaltered. Choose several handfuls of re-expressed predictor variables, which have the largest correlation coefficients with the unaltered dependent variable. Then proceed as usual, invoking the Bulging Rule for exploring the best re-expressions of the dependent variable and the predictor variables. If the dependent variable is re-expressed, then apply steps 1 or 2, above.

Meanwhile, there is an approach considered the most desirable for picking out the best re-expressed quantitative dependent variable; it is, however, neither practical nor easily assessable. It is outlined in Tukey and Mosteller's *Data Analysis and Regression* ("Graphical Fitting by Stages," pages 271–279). However, this approach is extremely tedious to perform manually as is required because there is no commercially available software for its calculations. Its inaccessibility has no consequence to the data analysts' quality of model, as the approach has not provided noticeable improvement over the procedure in step 3 for database applications where models are implemented at the decile-level.

Now that we have examined all the issues surrounding the quantitative dependent variable, I return to a discussion of re-expressing the predictor variables, starting with INCOME, and then AGE.

4.2.1.1 Re-expressing INCOME

I envision an underlying positively sloped straight line running through the ten points in the PROFIT-INCOME smooth plot in Figure 4.2, even though the smooth trace reveals four severe kinks. Based on the General Association Test with TS value of 6, which is *almost* equal to the cut-off score 7, as presented in Chapter 2, I conclude there is an *almost noticeable* straight-line relationship between PROFIT and INCOME. The correlation coefficient for the relationship is a reliable $r_{PROFIT, INCOME}$ of 0.763. Notwithstanding these indicators of straightness, the relationship could use some straightening; but, clearly, the Bulging Rule does not apply.

An alternative method for straightening data, especially characterized by nonlinearities, is the GenIQ procedure, a machine-learning genetic-based data mining method. As I extensively cover this method in Chapters 15 and 16, suffice it to say, I use GenIQ to re-express INCOME. The genetically evolved structure, which represents the re-expressed INCOME variable, labeled gINCOME, is defined in Equation (4.3):

$$gINCOME = sin(sin(sin(sin(INCOME) * INCOME) + log(INCOME) \quad (4.3)$$

The structure uses the nonlinear re-expressions of the trigonometric sine function (four times!) and the log (to base 10) function to loosen the "kinky" PROFIT-INCOME relationship. The relationship between PROFIT and INCOME (via gINCOME) has indeed been smoothed out, as the smooth

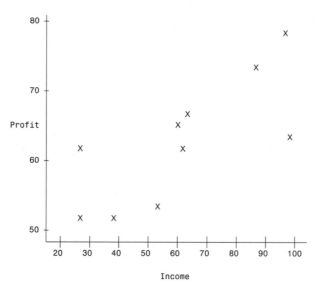

FIGURE 4.2
Plot of PROFIT and INCOME

trace reveals no serious kinks in Figure 4.3. Based on TS equal 6, which again is almost equal to the cutoff score of 7, I conclude there is an almost noticeable straight-line PROFIT–gINCOME relationship, a nonrandom scatter about an underlying positively sloped straight-line. The correlation coefficient for the re-expressed relationship is a reliable $r_{\text{PROFIT, gINCOME}}$ of 0.894.

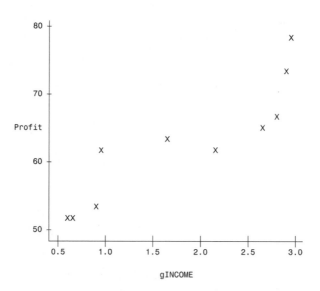

FIGURE 4.3
Plot of PROFIT and gINCOME

Visually, the effectiveness of the GenIQ procedure in straightening the data is obvious: the sharp peaks and valleys in the original PROFIT smooth plots vs. the smooth wave of the re-expressed smooth plot. Quantitatively, the gINCOME-based relationship represents a noticeable improvement of 7.24% (= (0.894 - 0.763)/0.763) increase in correlation coefficient "points" over the INCOME-based relationship.

Two points of note: recall that I previously invoked the statistical factoid that states a dollar-unit variable is often re-expressed with the log function. Thus, it is not surprising that the genetically evolved structure gINCOME uses the log function. With respect to logging the PROFIT variable, I concede that PROFIT could not benefit from a log re-expression, no doubt due to the "mini" in the dataset (i.e., the small size of the data), so I chose to work with PROFIT, not log of PROFIT, for the sake of simplicity (another EDA mandate, even for instructional purposes).

4.2.1.2 Re-expressing AGE

The stormy scatter of the ten-paired (PROFIT, AGE) points in the smooth plot in Figure 4.4 is an exemplary plot of a nonrelationship between two variables. Not surprisingly, the TS value of 3 indicates there is no noticeable PROFIT-AGE relationship. Senselessly, I calculate the correlation coefficient for this nonexistent linear relationship: $r_{PROFIT, AGE}$ equals −0.172, which is clearly not meaningful. Clearly, the Bulging Rule does not apply.

I use GenIQ to re-express AGE, labeled gAGE. The genetically based structure is defined in Equation (4.4):

$$gAGE = sin(tan(tan(2*AGE) + cos(tan(2*AGE)))) \qquad (4.4)$$

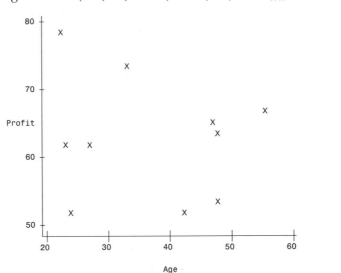

FIGURE 4.4
Plot of PROFIT and AGE

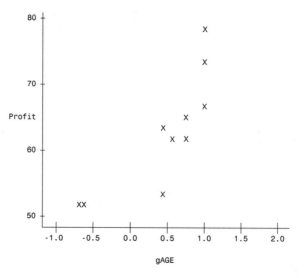

FIGURE 4.5
Plot of PROFIT and gAGE

The structure uses the nonlinear re-expressions of the trigonometric sine, cosine and tangent functions to calm the stormy-nonlinear relationship. The relationship between PROFIT and AGE (via gAGE) has indeed been smoothed out, as the smooth trace reveals in Figure 4.5. There is an almost noticeable PROFIT-gAGE relationship with TS = 6, which favorably compares to the original TS of 3! The re-expressed relationship admittedly does not portray an exemplary straight line, but given its stormy origin, I see a beautiful positively sloped ray, not very straight, but trying to shine through. I consider the corresponding correlation coefficient r $_{PROFIT, gAGE}$ value of 0.819 as reliable and remarkable.

Visually, the effectiveness of the GenIQ procedure in straightening the data is obvious: the abrupt spikes in the original smooth plot of PROFIT and AGE vs. the rising counter-clockwise *wave* of the second smooth plot of PROFIT and gAGE. With enthusiasm and without quantitative restraint, the gAGE-based relationship represents a noticeable improvement — a whopping 376.2% (= (0.819 − 0.172)/0.172; disregarding the sign) improvement in correlation coefficient points over the AGE-based relationship. Since the original correlation coefficient is meaningless, the improvement percent is also meaningless.

4.2.2   Smooth Predicted vs. Actual Plot

For a closer look at the detail of the strength (or weakness) of the gINCOME and gAGE structures, I construct the corresponding PROFIT smooth predicted vs. actual plots. The scatter about the 45-degree lines in the smooth plots for both gINCOME and gAGE in Figures 4.6 and 4.7, respectively,

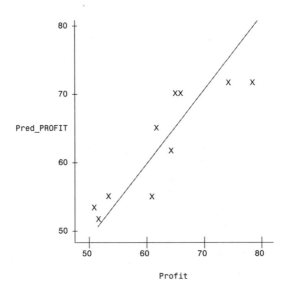

FIGURE 4.6
Smooth PROFIT Predicted vs. Actual based on gINCOME

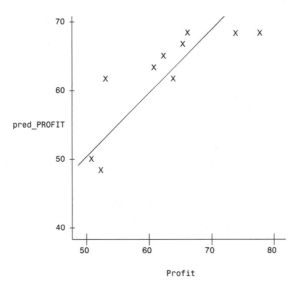

FIGURE 4.7
Smooth PROFIT Predicted vs. Actual based on gAGE

indicate a reasonable level of certainty in the reliability of the structures. In other words, both gINCOME and gAGE should be an important variable for predicting PROFIT. The correlations between gINCOME-based predicted and actual smooth PROFIT values, and between gAGE-based predicted and actual smooth PROFIT values, have $r_{sm.PROFIT, sm.gINCOME}$ and $r_{sm.PROFIT, sm.gAGE}$

values equal to 0.894, and 0.819, respectively. (Why are these r values equal to r PROFIT, INCOME and r PROFIT, AGE, respectively?)

4.2.3 Assessing the Importance of Variables

As in the correlating section of Chapter 3, the classical approach of assessing the statistical significance of a variable for model inclusion is the well-known null hypothesis-significance testing procedure, which is based on the reduction in prediction error (actual profit minus predicted profit) associated with the variable in question. The only difference between the discussion of the logistic regression in Chapter 3 is the apparatus used. The statistical apparatus of the formal testing procedure for ordinary regression consists of: the sum of squares (total; due to regression; due to error), the F statistic, degrees of freedom, and the p-value. The procedure uses the apparatus within a theoretical framework with weighty and untenable assumptions which, from a purist's point of view, can cast doubt on findings of statistical significance. Even if findings of statistical significance are accepted as correct, it may not be of practical importance or have *noticeable* value to the study at hand. For the data analyst with a pragmatist slant, the limitations and lack of scalability of the classical system of variable assessment cannot be overlooked, especially within big data settings. In contrast, the data mining approach uses the F statistic, and R-squared and degrees of freedom in an informal data-guided search for variables that suggest a *noticeable* reduction in prediction error. Note that the informality of the data mining approach calls for suitable change in terminology, from declaring a result as statistically significant to worthy of notice or *noticeably important*.

4.2.3.1 Defining the F Statistic and R-squared

In data mining, the assessment of the importance of a subset of variables for predicting profit involves the notion of a *noticeable* reduction in prediction error due to the subset of variables. It is based on the F statistic, R-squared and degrees of freedom (df), which are always reported in the ordinary regression output. For the sake of reference, I provide their definitions and relationship with each other in Equations (4.5), (4.6) and (4.7) below.

$$F = \frac{\text{Sum of squares due to regression} / \text{df due to regression model}}{\text{Sum of squares due to error} / \text{df due to error in regression model}} \quad (4.5)$$

$$R\text{-squared} = \frac{\text{Sum of squares due to regression}}{\text{Total Sum of squares}} \quad (4.6)$$

$$F = \frac{R\text{-squared} / \text{number of variables in model}}{(1 - R\text{-squared}) / (\text{sample size} - \text{number of variables in model} - 1)} \quad (4.7)$$

For the sake of completion, I provide an additional statistic: the adjusted R-squared. R-squared is affected, among other things, by the ratio of the number of predictor variables in the model to the size of the sample. The larger the ratio the greater the overestimation of R-squared. Thus, the adjusted R-squared as defined in Equation (4.8) is not particularly useful in big data settings.

Adjusted R-squared =

$$1 - (1 - R\text{-squared}) \frac{(\text{sample size} - 1)}{(\text{sample size} - \text{number of variables in model} - 1)} \quad (4.8)$$

In the sections below, I detail the decision rules for three scenarios for assessing the importance of variables, i.e., the likelihood the variables have *some* predictive power. In brief, the larger the F statistic, R-squared and adjusted R-squared values, the more important the variables are in predicting profit.

4.2.3.2 Importance of a Single Variable

If X is the only variable considered for inclusion into the model, the decision rule for declaring X an important predictor variable in predicting profit is: if the F value due to X is greater than the *standard F value 4*, then X is an important predictor variable and should be considered for inclusion in the model. Note that the decision rule only indicates that the variable has some importance, not how much importance. The decision rule implies that a variable with a greater F value has a greater *likelihood of some importance* than a variable with a smaller F value, not that it has greater importance.

4.2.3.3 Importance of a Subset of Variables

When subset A consisting of k variables is the only subset considered for model inclusion, the decision rule for declaring subset A important in predicting profit is as follows: if the average F value per number of variables (the degrees of freedom) in the subset A — F/df or F/k — is greater than standard F value 4, then subset A is an important subset of predictor variable and should be considered for inclusion in the model. As before, the decision rule only indicates that the subset has some importance, not how much importance.

4.2.3.4 Comparing the Importance of Different Subsets of Variables

Let subsets A and B consist of k and p variables, respectively. The number of variables in each subset does not have to be equal. If they are equal, then all but one variable can be the same in both subsets. Let F(k) and F(p) be the F values corresponding to the models with subset A and B, respectively.

The decision rule for declaring which of the two subsets is more important (greater likelihood of some predictive power) in predicting profit is:

1. If $F(k)/k$ greater than $F(p)/p$, then subset $A(k)$ is the more important predictor variable subset; otherwise, $B(p)$ is the more important subset.

2. If $F(k)/k$ and $F(p)/p$ are equal or have comparable values then both subsets are to be regarded tentatively of comparable importance. The data analyst should consider additional indicators to assist in the decision about which subset is better. It clearly follows from the decision rule that the model defined by the more important subset is the better model. (Of course, the rule assumes that F/k and F/p are greater than the standard F value 4.)

Equivalently, the decision rule can use either the R-squared or adjusted R-squared in place of the F/k and F/p. The R-squared statistic is a friendly concept, in that its values serve as an indicator of "the percent of variation explained by the model."

4.3   Important Variables for Mini Case Study

I perform two ordinary regressions, regressing PROFIT on gINCOME, and on gAGE; the outputs are in Tables 4.5 and 4.6, respectively. The F values

TABLE 4.5

OLS Output: PROFIT with gINCOME

Source	df	Sum of Squares	Mean Square	F Value	Pr > F
Model	1	582.1000	582.1000	31.83	0.0005
Error	8	146.3000	18.2875		
Corrected Total	9	728.4000			
		Root MSE	4.2764	R-Square	0.7991
		Dependent Mean	62.6000	Adj R-Sq	0.7740
		Coeff Var	6.8313		

Variable	df	Parameter Estimate	Standard Error	t Value	Pr > |t|
INTERCEPT	1	47.6432	2.9760	16.01	<.0001
gINCOME	1	8.1972	1.4529	5.64	0.0005

TABLE 4.6

OLS Output: PROFIT with gAGE

Source	df	Sum of Squares	Mean Square	F Value	Pr > F
Model	1	488.4073	488.4073	16.28	0.0038
Error	8	239.9927	29.9991		
Corrected Total	9	728.4000			
		Root MSE	5.4771	R-Square	0.6705
		Dependent Mean	62.6000	Adj R-Sq	0.6293
		Coeff Var	8.7494		

Variable	df	Parameter Estimate	Standard Error	t Value	Pr > |t|
INTERCEPT	1	57.2114	2.1871	26.16	< .0001
gAGE	1	11.7116	2.9025	4.03	0.0038

are 31.83 and 16.28, respectively, which are greater than the standard F value 4. Thus, both gINCOME and gAGE are declared important predictor variables of PROFIT.

4.3.1 Relative Importance of the Variables

Chapter 3 contains a similar subheading for which there is only a minor variation with respect to the statistic used. The *t statistic* as posted in ordinary regression output can serve as an indicator of a variable's relative importance, and for selecting the best subset, which is discussed in the next section.

4.3.2 Selecting the Best Subset

The decision rules for finding the best subset of important variables are nearly the same as those discussed in Chapter 3. Refer to that chapter's subheading, Section 3.11.1 *Selecting the Best Subset*. Point 1 remains the same for this discussion. However, the second and third points change, as follows:

1. Select an initial subset of important variables.
2. For the variables in the initial subset, generate smooth plots and straighten the variables as required. The most noticeable handfuls of original and re-expressed variables form the starter subset.

3. Perform the preliminary ordinary regression on the starter subset. Delete one or two variables with absolute t statistic values less than the *t cut-off value* 2 from the model. This results in the first incipient subset of important variables. Note the changes to points four, five and six with respect to this chapter's topic.

4. Perform another ordinary regression on the incipient subset. Delete one or two variables with t values less than the t cut-off value 2 from the model. The data analyst can create an illusion of important variables appearing and disappearing with the deletion of different variables. The remainder of the discussion in Chapter 3 remains the same.

5. Repeat step 4 until all retained predictor variables have comparable t values. This step often results in different subsets, as the data analyst deletes judicially different pairings of variables.

6. Declare the best subset by comparing the relative importance of the different subsets using decision rule in Section 4.2.3.4.

4.4 Best Subset of Variable for Case Study

I build a preliminary model by regressing PROFIT on gINCOME and gAGE; the output is in Table 4.7. The two-variable subset has an F/df value of 7.725 (= 15.45/2), which is greater than the standard F value 4. But the t value for gAGE is 0.78 less than the t cut-off value (see bottom section of Table

TABLE 4.7

OLS Output: PROFIT with gINCOME and gAGE

Source	df	Sum of Squares	Mean Square	F Value	Pr > F
Model	2	593.8646	296.9323	15.45	0.0027
Error	7	134.5354	19.2193		
Corrected Total	9	728.4000			
		Root MSE	4.3840	R-Squared	0.8153
		Dependent Mean	62.6000	Adj R-Sq	0.7625
		Coeff Var	7.0032		

Variable	df	Parameter Estimate	Standard Error	t Value	Pr > |t|
INTERCEPT	1	49.3807	3.7736	13.09	< .0001
gINCOME	1	6.4037	2.7338	2.34	0.0517
gAGE	1	3.3361	4.2640	0.78	0.4596

4.7). If I follow step 4 above, then I would have to delete gAGE, yielding a simple regression model with the lowly, abeit straight, predictor variable gINCOME. By the way, the adjusted R-squared is 0.7625 (after all, the entire mini sample is not big).

Before I dismiss the two-variable (gINCOME, gAGE) model, I construct the smooth residual plot in Figure 4.8 to determine the quality of the predictions of model. The smooth residual plot is declared to be equivalent to the null plot based General Association Test (TS = 5). Thus, the overall quality of the predictions is considered good. That is, on average, the predicted PROFIT is equal to the actual PROFIT. The descriptive statistics for the smooth residual plot are: for the smooth residual, the minimum and maximum values and range are −4.567 and 6.508, and 11.075, respectively; the standard deviation of the smooth residuals is 3.866.

4.4.1 PROFIT Model with gINCOME and AGE

With a vigilance in explaining the unexpected, I suspect the reason for gAGE's relative nonimportance (i.e., gAGE is not important in the presence of gINCOME) is gAGE's strong correlation with gINCOME: $r_{\text{gINCOME, gAGE}}$ = 0.839. This supports my contention, but does not confirm it.

In view of the foregoing, I build another two-variable model regressing PROFIT on gINCOME and AGE; the output is in Table 4.8. The (gINCOME, AGE) subset has an F/df value of 12.08 (= 24.15/2), which is greater than the standard F value 4. Statistic-happy, I see the t values for both variables are greater than t cut-off value 2. I cannot overlook that fact that the raw variable AGE, which by itself is not important, now has relative importance in the presence of gINCOME! (More about this "phenomenon" at the end of the chapter.) Thus, the evidence is that gINCOME and AGE is a better subset than the original (gINCOME, gAGE) subset. By the way, the adjusted

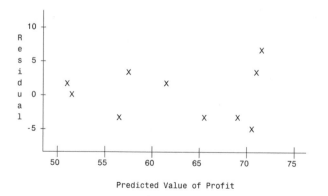

FIGURE 4.8
Smooth Residual Plot for (gINCOME,gAGE) Model

TABLE 4.8

OLS Output: PROFIT with gINCOME and AGE

Source	df	Sum of Squares	Mean Square	F Value	Pr > F
Model	2	636.2031	318.1016	24.15	0.0007
Error	7	92.1969	13.1710		
Corrected Total	9	728.4000			
		Root MSE	3.6291	R-Squared	0.8734
		Dependent Mean	62.6000	Adj R-Sq	0.8373
		Coeff Var	5.79742		

Variable	df	Parameter Estimate	Standard Error	t Value	Pr > |t|
INTERCEPT	1	54.4422	4.1991	12.97	< .0001
gINCOME	1	8.4756	1.2407	6.83	0.0002
AGE	1	-0.1980	0.0977	-2.03	0.0823

R-squared is 0.8373 representing a 9.81% (= (0.8373 − 0.7625)/0.7625) improvement in "adjusted R-squared" points over the original-variable adjusted R-squared.

The smooth residual plot for the (gINCOME, AGE) model in Figure 4.9 is declared to be equivalent to the null plot-based General Association Test with TS = 4. Thus, there is indication that the overall quality of the predictions is good. The descriptive statistics for the two-variables smooth residual plot are: for the smooth residual, the minimum and maximum values and range are −5.527 and 4.915, and 10.442; respectively, the standard deviation of the smooth residual is 3.200.

To obtain further indication of the quality of the model's predictions I construct the smooth actual vs. predicted plot for the (gINCOME, AGE) model in Figure 4.10. The smooth plot is acceptable with minimal scatter of the ten smooth points about the 45-degree line. The correlation between smooth actual vs. predicted PROFIT based on gINCOME and AGE has a $r_{sm.gINCOME, sm.AGE}$ value of 0.93. (This value is the square root of R-squared for the model. Why?)

For a point of comparison, I construct the smooth actual vs. predicted plot for the (gINCOME, gAGE) model in Figure 4.11. The smooth plot has acceptable with minimal scatter of the ten smooth points about the 45-degree lines, with a noted exception some wild scatter for PROFIT values greater than $65. The correlation between smooth actual vs. predicted PROFIT based on gINCOME and gAGE has a $r_{sm.gINCOME, sm.gAGE}$ value of 0.90. (This value is the square root of R-squared for the model. Why?)

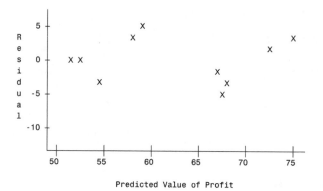

FIGURE 4.9
Smooth Residual Plot for (gINCOME, AGE) Model

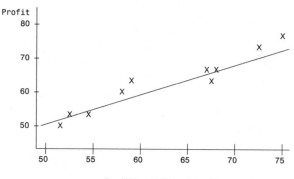

FIGURE 4.10
Smooth Actual vs. Predicted Plot for (gINCOME, AGE) Model

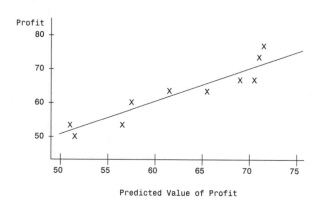

FIGURE 4.11
Smooth Actual vs. Predicted for (gINCOME, gAGE) Model

4.4.2 Best PROFIT Model

To decide which of the two PROFIT models is better, I put their vital statistics of the preceding analyses in Table 4.9. Based on the consensus of a committee of one, I prefer the gINCOME-AGE model over the gINCOME-gAGE model, because there are noticeable indications that the former model's predictions are better. The former model offers a 9.8% increase in the adjusted R-squared (less bias), a 17.2% decrease in the smooth residual standard deviation (more stable), and a 5.7% decrease in the smooth residual range (more stable).

4.5 Suppressor Variable AGE

A variable whose behavior is like that of AGE — poorly correlated with the dependent variable Y, but becomes important by its inclusion in a model for predicting Y — is known as a *suppressor variable*. [1, 2] The consequence of a suppressor variable is that it increases the model's R-squared.

I explain the behavior of the suppressor variable within the context of the mini case study. The presence of AGE in the model removes or suppresses the information (variance) in gINCOME that is *not related* to the variance in PROFIT, i.e., AGE suppresses the unreliable "noise" in gINCOME. This renders the AGE-adjusted variance in gINCOME more reliable or potent for predicting PROFIT.

I analyze the paired correlations among the three variables to clarify exactly what AGE is doing. Recall that squaring the correlation coefficient represents the "shared variance" between the two variables under consideration. The paired bits of information are in Table 4.10. I know from the prior analysis that PROFIT and AGE have no noticeable relationship; their shared variance of 3% confirms this. I also know that PROFIT and gINCOME do have a noticeable relationship; their shared variance of 80% confirms this as well. In the presence of AGE, the relationship between PROFIT and gINCOME, specifically, the relationship between PROFIT and gINCOME-adjusted-for-AGE has a shared variance of 87%. This represents an improvement of 8.75% (= (0.87 – 0.80/0.80) in shared variance. This "new" variance is now available for predicting PROFIT, increasing the R-squared (from 79.91 to 87.34%).

It is a pleasant surprise in several ways that AGE turns out to be a suppressor variable. First, suppressor variables occur most often in big data settings, not often with small data, and are truly unexpected with mini data. Second, they serve as object lessons for the EDA paradigm: dig, dig, dig into the data and you will find gold, or some reward for your effort. Third, it is a small reminder of a big issue: the data analyst must not rely solely on predictor variables that are highly correlated with the dependent variable, but also consider the poorly correlated predictor variables, as they are a great source of latent predictive importance.

TABLE 4.9

Comparison of Vitals Statistics for Two PROFIT Models

Model	Predictor Variables	F value	t value	Smooth Residual		StdDev	Adj. R-square
				Range	StdDev		
First	gINCOME, gAGE	Greater than cut-off value	Greater than cut-off value for only gINCOME	11.075	3.866		0.7625
Second	gINCOME, AGE	Greater than cut-off value	Greater than cut-off value for both variables	10.442	3.200		0.8373
Indication	Improvement of 2nd model over 1st model	NA	Because t value for gAGE is less than t cut-off value, gAGE contributes "noise" in model, as evidenced in range, stdDev and adj. R-sq.	–5.7%	–17.2%		9.8%

TABLE 4.10

Comparison of Pairwise Correlations Among PROFIT, AGE and gINCOME

Correlation Pair	Correlation Coefficient	Shared Variance
PROFIT and AGE	−0.172	3%
PROFIT and gINCOME	0.894	80%
PROFIT and AGE in the presence of gINCOME	−0.608	37%
PROFIT and gINCOME in the presence of AGE	0.933	87%

4.6 Summary

The ordinary regression model is presented as the workhorse of database profit modeling, as it has been in steady use for almost two hundred years. As such, I illustrated in an orderly and detailed way the essentials of ordinary regression. Moreover, I showed the enduring usefulness of this popular analysis and modeling technique, as it works well within today's EDA/data mining paradigm.

I first illustrated the rudiments of the ordinary regression model by discussing the SAS program code for building and scoring an ordinary regression model. The code is a welcome addition to the techniques used by data analysts working on predicting a quantitative dependent variable.

Then, I discussed ordinary regression modeling with a mini data. I used this extremely small data to not only make the discussion of the data mining techniques tractable, but to emphasize two aspects of data mining. First, data mining techniques of great service should work as well with big data as with small data, as explicitly stated in the definition of data mining in Chapter 1. Second, every fruitful effort of data mining on small data is evidence that big data are not always necessary to uncovered structure in the data. This evidence is in keeping with the EDA philosophy that the data analyst should work from simplicity until indicators emerge to go further: if predictions are not acceptable, then increase data size. The data mining techniques discussed are those introduced in logistic regression framework of Chapter 3, and carry over with minor modification to ordinary regression.

Before proceeding with the analysis of the mini case study, I supplemented the Bulging Rule when analysis involves ordinary regression. Unlike in logistic regression where the logit dependent variable cannot be re-expressed, in ordinary regression the quantitative dependent variable can be re-expressed. The Bulging Rule as introduced within the logistic regression framework in Chapter 3 can put several re-expressions of the quantitative dependent variable up for consideration. For such cases, I provided additional guidelines to the Bulging Rule, which will prove valuable.

1. If there is a small range of dependent-variable powers then, the best re-expressed dependent variable is the one with the noticeably largest correlation coefficient.

2. If there is a small range of dependent-variable powers and the correlation coefficient values are comparable, then the best re-expressed dependent-variable is defined by the average power among the dependent-variable powers.

3. When there is a large range of dependent-variable powers: choose several handfuls of re-expressed predictor variables, which have the largest correlation coefficients with the unaltered quantitative dependent variable. Then proceed as usual, invoking the Bulging Rule for exploring the best re-expressions of the dependent-variable and the predictor variables. If the dependent-variable is re-expressed, then apply steps 1 or 2, above.

With the mini dataset selected for regressing PROFIT on INCOME and AGE, I introduced the alternative method GenIQ, a machine learning genetic-based data mining method for straightening data. I illustrated the data mining procedure of smoothing and assessing the smooth with the General Association Test and determined that both predictor variables need straightening, but the Bulging Rule does not apply. The GenIQ method evolved reasonably straight relationships between PROFIT and each re-expressed predictor variables, gINCOME and gAGE, respectively. I generated the PROFIT smooth predictive vs. actual plots, which provided further indication that gINCOME and gAGE should be important variables for predicting PROFIT.

Continuing with the mini case study, I demonstrated an ordinary regression-specific data mining alternative approach to the classical method of assessing. (The data mining techniques discussed are those introduced in logistic regression framework of Chapter 3, and carry over with minor modification to ordinary regression.) This alternative approach involves the importance of individual predictor variables, as well as the importance of a subset of predictor variables and the relative importance of individual predictor variables, in addition to the goodness of model predictions. Additional methods that set out specific to ordinary regression included: selecting the best subset of predictor variables, and comparing the importance between two subsets of predictor variables.

Within my illustration of the case study is a pleasant surprise — the existence of a suppressor variable (AGE). A variable whose behavior is poorly correlated with the dependent variable, but becomes important by its inclusion in a model for predicting the dependent variable, is known as a suppressor variable. The consequence of a suppressor variable is that it increases the model's R-squared. Suppressor variables occur most often in big data settings, not often with small data, and are truly unexpected with mini data. Second, they serve as object lessons for the EDA paradigm: dig deeply into the data and you will find a reward for your effort. Third, it is a small reminder of a bigger issue — that the data analyst must not rely solely on predictor variables that are highly correlated with the dependent variable, but also should consider the poorly correlated predictor variables, as they are a great source of latent predictive importance.

References

1. Horst, P., The role of predictor variables which are independent of the criterion. *Social Science Research Bulletin*, 48, 431–436, 1941.
2. Conger, A.J., A revised definition for suppressor variables: A guide to their identification and interpretation, *Educational and Psychological Measurement*, 34, 35–46, 1974.

5

CHAID for Interpreting a Logistic Regression Model[1]

The logistic regression model is the standard technique for building a response model in database marketing. Its theory is well established and its estimation algorithm is available in all major statistical software packages. The literature on the theoretical aspects of logistic regression is large and rapidly growing. However, little attention is paid to the interpretation of logistic regression response model. The purpose of this chapter is to present a CHAID-based method for interpreting a logistic regression model, specifically to provide a complete assessment of the effects of the predictor variables on response.

5.1 Logistic Regression Model

I briefly state the definition of the logistic regression model. Let Y be a binary response (dependent) variable, which takes on "yes-no" values (typically coded 1–0, respectively), and X_1, X_2, ..., X_n be the predictor (independent) variables. The logistic regression model estimates the *logit of Y* — a log of the odds of an individual responding "yes" as defined in Equation (5.1), from which an individual's probability of responding "yes is" obtained in Equation (5.2):

$$\text{logit } Y = b_0 + b_1{}^*X_1 + b_2{}^*X_2 + ... + b_n{}^*X_n \tag{5.1}$$

$$\text{Prob}(Y = 1) = \frac{\exp(\text{Logit } Y)}{1 + \exp(\text{Logit } Y)} \tag{5.2}$$

---

[1] This chapter is based on an article with the same title in *Journal of Targeting, Measurement and Analysis for Marketing*, 6, 2, 1997. Used with permission.

An individual's predicted probability of responding "yes" is calculated by "plugging-in" the values of the predictor variables for that individual in the Equations (5.1) and (5.2). The bs are the logistic regression coefficients. The coefficient b_0 is referred to as the Intercept, and has no predictor variable with which it is multiplied.

The *odds ratio* is the traditional measure of assessing the effect of a predictor variable on the response variable (actually on the odds of response = 1) given that the other predictor variables are "held constant." The latter phrase implies that the odds ratio is the average effect of a predictor variable on response when the effects of the other predictor variables are "partialled out" of the relationship between response and the predictor variable. Thus, the odds ratio does not explicitly reflect the variation of the other predictor variables. The odds ratio for a predictor variable is obtained by exponentiating the coefficient of the predictor variable. That is, the odds ratio for X_i equals $exp(b_i)$, where exp is the exponential function, and b_i is the coefficient of X_i.

5.2   Database Marketing Response Model Case Study

A woodworker's tool supplier, who wants to increase response to her catalog in an upcoming campaign, needs a model for generating a list of her most responsive customers. The response model, which is built on a sample drawn from a recent catolog mailing with a 2.35% response rate, is defined by the following variables:

1. The response variable is RESPONSE, which indicates whether a customer made a purchase (yes = 1, no = 0) to the recent catalog mailing.

2. The predictor variables are: a) CUST_AGE — the customer age in years; b) LOG_LIFE — the log of total purchases in dollars since the customer's first purchase, that is, the log of lifetime dollars; and c) PRIOR_BY — the dummy variable indicating whether a purchase was made in prior three months to the recent catalog mailing (yes = 1, no = 0).

The logistic regression analysis on RESPONSE with the three predictor variable produces the output in Table 5.1. The "Estimate" column contains the logistic regression coefficients, from which the RESPONSE Model is defined in Equation (5.3).

$$\text{logit RESPONSE} = -8.43 + 0.02*\text{CUST\_AGE} + 0.74*\text{LOG\_LIFE} + 0.82*\text{PRIOR\_BY} \tag{5.3}$$

TABLE 5.1

Logistic Regression Output

Variable		Parameter Estimate	Standard Error	Wald Chi-square	Pr> Chi-square	Odds Ratio
INTERCEPT	1	−8.4349	0.0854	9760.7175	1.E + 00	
CUST_AGE	1	0.0223	0.0004	2967.8450	1.E + 00	1.023
LOG_LIFE	1	0.7431	0.0191	1512.4483	1.E + 00	2.102
PRIOR_BY	1	0.8237	0.0186	1962.4750	1.E + 00	2.279

5.2.1 Odds Ratio

The following discussion illustrates two weaknesses in the odds ratio. One, the odds ratio is in terms of odds-of-responding-yes units, which all but the mathematically adept feel comfortable with. Two, the odds ratio provides a "static" assessment of a predictor variable's effect, as its value is constant regardless of the relationship among the other predictor variables. The odds ratio is part of the standard output of the logistic regression analysis. For the case study, the odds ratio is in the last column in Table 5.1.

1. For PRIOR_BY with coefficient value 0.8237, the odds ratio is 2.279 (= exp(0.8237)). This means that for a unit increase in PRIOR_BY, i.e., going from 0 to 1, the odds (of responding yes) for an individual who has made a purchase within the prior three months is 2.279 times the odds of an individual who has <u>not</u> made a purchase within the prior three months — given CUST_AGE and LOG_LIFE are held constant.[2]

2. CUST_AGE has an odds ratio of 1.023. This indicates that for every one year increase in a customer's age, the odds increase by 2.3% — given PRIOR_BY and LOG_LIFE are held constant.

3. LOG_LIFE has an odds ratio of 2.102, which indicates that for every one log-lifetime-dollar unit increase, the odds increase by 110.2% — given PRIOR_BY and CUST_AGE are held constant.

The proposed CHAID-based method supplements the odds ratio for interpreting the effects of a predictor on response. It provides treelike displays in terms of nonthreatening probability units, everyday values ranging from 0% to 100%. Moreover, the CHAID-based graphics provide a complete assessment of a predictor variable's effect on response. It brings forth the simple, unconditional relationship between a given predictor variable and response, as well as the conditional relationship between a given predictor

---

[2] For given values of CUST_AGE and LOG_LIFE, say, a and b, respectively, the odds ratio for PRIOR_BY is defined as:

$$\frac{\text{odds (PRIOR\_BY} = 1 \text{ given CUST\_AGE} = a \text{ and LOG\_LIFE} = b)}{\text{odds (PRIOR\_BY} = 0 \text{ given CUST\_AGE} = a \text{ and LOG\_LIFE} = b)}$$

variable and response shaped by the relationships between the other X's and a given predictor variable, and the other X's and response.

5.3   CHAID

Briefly stated, CHAID is a technique that recursively partitions a population into separate and distinct sub-populations or segments such that the variation of the dependent variable is minimized within the segments, and maximized among the segments. A CHAID analysis results in a treelike diagram, commonly called a CHAID tree. CHAID was originally developed as a method of finding "combination" or interaction variables. In database marketing today, CHAID primarily serves as a market segmentation technique. The use of CHAID in the proposed application for interpreting a logistic regression model is possible because of CHAID's salient data mining features. CHAID is eminently good in uncovering structure, in this application, within the conditional and unconditional relationships among response and predictor variables. Moreover, CHAID is excellent in graphically displaying multivariable relationships; its tree output is very easy to read and interpret.

It is worth emphasizing that in this application CHAID is not used as an alternative method for analyzing or modeling the data at hand. CHAID is used as a visual aid for depicting the "statistical" mechanics of a logistic regression model, such as how the variables of the model work together in contributing to the predicted probability of response. It is assumed that the response model is already built by any method, not necessarily logistic regression. As such, the proposed method can enhance the interpretation of the predictor variables of any model built by statistical or machine learning techniques.

5.3.1   Proposed CHAID-Based Method

To perform an ordinary CHAID analysis, the data analyst is required to select both the response variable and a set of predictor variables. For the proposed CHAID-based method, the already-built response model's *estimated probability of response* is selected as the CHAID-response variable. The CHAID-set of predictor variables consists of the predictor variables that defined the response model in their original units, not in re-expressed units (if re-expressing was necessary). Re-expressed variables are invariably in units that hinder the interpretation of the CHAID-based analysis, and, consequently, the logistic regression model. Moreover, to facilitate the analysis, the continuous predictor variables are categorized into meaningful intervals based on content domain of the problem under study.

For the case study, the CHAID-response variable is the estimated probability of RESPONSE, called PROB_EST, which is obtained from Equations (5.2) and (5.3) and defined in Equation (5.4):

$$\text{PROB\_EST} = \frac{\exp(-8.43 + 0.02*\text{CUST\_AGE} + 0.74*\text{LOG\_LIFE} + 0.82*\text{PRIOR\_BY})}{1 + \exp(-8.43 + 0.02*\text{CUST\_AGE} + 0.74*\text{LOG\_LIFE} + 0.82*\text{PRIOR\_BY})}$$

(5.4)

The CHAID-set of predictor variables are: CUST_AGE, categorized into two classes, PRIOR_BY, and the original variable LIFETIME DOLLARS (log-lifetime-dollars units are hard to understand!), categorized into three classes. The woodworker views her customers in terms of the following classes:

1. The two CUST_AGE classes are: less than 35 years, and 35 years & up. CHAID, which uses bracket and parenthesis symbols in its display of intervals, denotes the two customer age intervals as: [18, 35) and [35, 93], respectively. CHAID defines intervals as closed interval and left-closed/right-open interval. The former is denoted by [a,b] indicating all values between and including a & b. The latter is denoted by [a,b) indicating all values greater than/equal to a, and less than b. Minimum and maximum ages in the sample are 18 and 93, respectively.

2. The three LIFETIME DOLLARS classes are: less than $15,000; $15,001 to $29,999; and equal to or greater than $30,000. CHAID denotes the three lifetime dollar intervals as: [12, 1500), [1500, 30000) and [30000, 675014], respectively. Minimum and maximum lifetime dollars in the sample are $12 and $675,014, respectively.

The CHAID trees, in Figures 5.1 to 5.3, based on the PROB_EST variable for the three predictor variables are read as follows:

1. All CHAID trees have a top box (root node), which represents the sample under study: sample size and response rate. For the proposed CHAID application, the top box reports the sample size and the average estimated probability (AEP) of response. For the case study, the sample size is 858,963, and AEP of RESPONSE is 0.0235.[3]

2. The CHAID tree for PRIOR_BY is in Figure 5.1: the left-leaf node represents a segment (size 333,408) defined by PRIOR_BY = no. These customers have not made a purchase in the prior three months; their AEP of RESPONSE is 0.0112. The right-leaf node represents a segment (size 525,555) defined by PRIOR_BY = yes. These customers have made a purchase in the prior three months; their AEP of RESPONSE is 0.0312.

---

[3] The average estimated probability or response rate is always equal to the true response rate.

FIGURE 5.1
CHAID Tree for PRIOR_BY

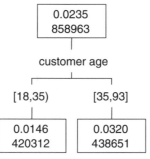

FIGURE 5.2
CHAID Tree for CUST_AGE

3. The CHAID tree for CUST_AGE is in Figure 5.2: the left-leaf node represents a segment (size 420,312) defined by customers whose ages are in the interval [18, 35); their AEP of RESPONSE is 0.0146. The right-leaf node represents a segment (size 438,651) defined by customers whose ages are in the interval [35,93]; their AEP of RESPONSE is 0.0320.

4. The CHAID tree for LIFETIME DOLLARS is in Figure 5.3: the left-leaf node represents a segment (size 20,072) defined by customers whose lifetime times are in the interval [12, 1500); their AEP of RESPONSE is 0.0065. The middle-leaf node represents a segment (size 613,965) defined by customers whose lifetime dollars are in the interval [1500, 30000); their AEP of RESPONSE is 0.0204. The left-leaf node represents a segment (size 224,926) defined by customers whose lifetime dollars are in the interval [30000, 675014]; their AEP of RESPONSE is 0.0332.

At this point, the single predictor variable CHAID tree shows the effect of the predictor variable on RESPONSE. Reading the leaf nodes from left to right, it is clearly revealed how RESPONSE changes as the predictor

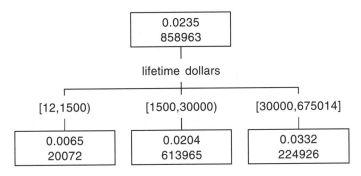

FIGURE 5.3
CHAID Tree for LIFETIME DOLLARS

variable's values increase. Although the single predictor variable CHAID tree is easy to interpret with probabilities units, it is still like the odds ratio, in that it does not reveal the effects of a predictor variable with respect to the variation of the other predictor variables in the model. For a complete visual display of the effect of a predictor variable on response accounting for the presence of other predictor variables, a *multivariable CHAID tree* is required.

5.4 Multivariable CHAID Trees

The multivariable CHAID tree in Figure 5.4 shows the effects of LIFETIME DOLLARS on RESPONSE with respect to the variation of CUST_AGE and PRIOR_BY = *no*. The LIFETIME DOLLARS-PRIOR_BY = no CHAID tree is read and interpreted as follows:

1. The root node represents the sample (size 858,963) with AEP of RESPONSE, 0.0235.

2. The tree has six *branches*, which are defined by the intersection of the CUST_AGE and LIFETIME DOLLARS intervals/nodes. Branches are read from an end-leaf node (bottom box) upward to and through intermediate-leaf nodes, stopping at the first-level leaf node below the root node.

3. Reading the tree starting at the bottom of the multivariable CHAID tree in Figure 5.4 from left to right: the left-most branch #1 is defined by LIFETIME DOLLARS = [12, 1500) and CUST_AGE = [18, 35) and PRIOR_BY = no;

   the second left-most branch #2 is defined by LIFETIME DOLLARS = [1500, 30000) and CUST_AGE = [18, 35) and PRIOR_BY = no;

   branches #3 to #5 are similarly defined;

   the right-most branch #6 is defined by LIFETIME DOLLARS = [30000, 675014] and CUST_AGE = [35, 93] and PRIOR_BY = no.

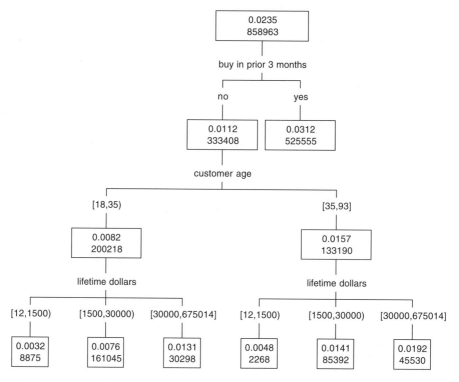

FIGURE 5.4
Multivariable CHAID Tree for Effects of LIFETIME DOLLARS, accounting for CUST_AGE and PRIOR_BY = no

4. Focusing on the three left-most branches #1 to #3, as LIFETIME DOLLARS increase, its effect on RESPONSE is gleaned from the multivariable CHAID tree: AEP of RESPONSE goes from 0.0032 to 0.0076 to 0.0131 for customers whose ages are in the interval [18, 35) *and* have not purchased in prior three months.

5. Focusing on the three right-most branches #4 to #6, as LIFETIME DOLLARS increase, its effect on RESPONSE is gleaned from the multivariable CHAID tree in Figure 5.4: AEP of RESPONSE ranges from 0.0048 to 0.0141 to 0.0192 for customers whose ages are in the interval [35, 93] *and* have not purchased in prior three months.

The multivariable CHAID tree in Figure 5.5 shows the effect of LIFETIME DOLLARS on RESPONSE with respect to the variation of CUST_AGE and PRIOR_BY = *yes*. Briefly, the LIFETIME DOLLARS-PRIOR_BY = yes CHAID tree is interpreted as follows:

1. Focusing on the three left-most branches #1 to #3, as LIFETIME DOLLARS increase, its effect on RESPONSE is gleaned from the

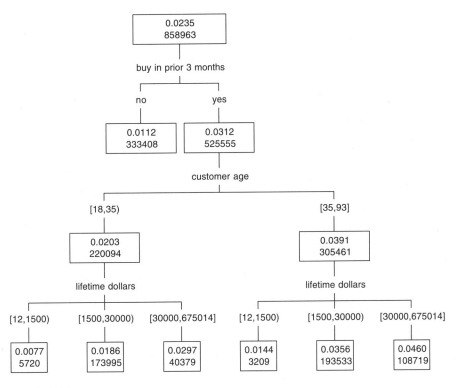

FIGURE 5.5
Multivariable CHAID Tree for Effects of LIFETIME DOLLARS, accounting for CUST_AGE and PRIOR_BY = yes

multivariable CHAID tree: AEP of RESPONSE goes from 0.0077 to 0.0186 to 0.0297 for customers whose ages are in the interval [35, 93] *and* have purchased in prior three months.

2. Focusing on the three right-most branches #4 to #6, as LIFETIME DOLLARS increase, its effect on RESPONSE is gleaned from the multivariable CHAID tree: AEP of RESPONSE goes from 0.0144 to 0.0356 to 0.0460 for customers whose ages are in the interval [35, 93] *and* have purchased in prior three months.

The multivariable CHAID trees for the effects of PRIOR_BY on RESPONSE with respect to the variation of CUST_AGE = [18, 35] and LIFETIME DOLLARS, and with respect to the variation of CUST_AGE = [35, 93] and LIFETIME DOLLARS are in Figures 5.6 and 5.7, respectively. They are similarly read and interpreted in the same manner as the LIFETIME DOLLARS multivariable CHAID trees in Figures 5.4 and 5.5.

The multivariable CHAID trees for the effects of CUST_AGE on RESPONSE with respect to the variation of PRIOR_BY = *no* and LIFETIME DOLLARS, and with respect to the variation of PRIOR_BY = *yes* and

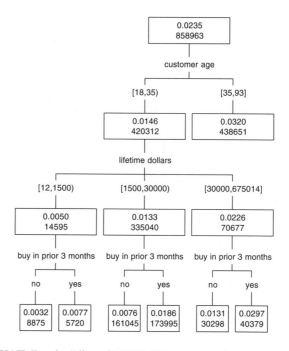

FIGURE 5.6

Multivariable CHAID Tree for Effect of PRIOR_BY, accounting for LIFETIME DOLLARS and CUST_AGE = [18, 35)

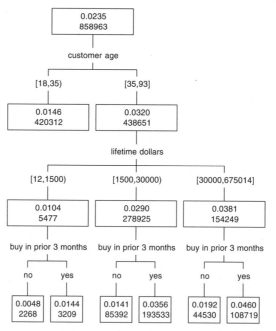

FIGURE 5.7

Multivariable CHAID Tree for Effect of PRIOR_BY, accounting for LIFETIME DOLLARS and CUST_AGE = [35, 93]

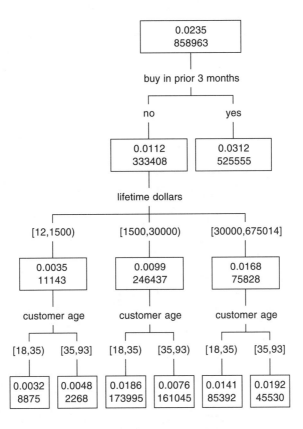

FIGURE 5.8
Multivariable CHAID Tree for Effect of CUST_AGE, accounting for LIFETIME DOLLARS and
PRIOR_BY = no

LIFETIME DOLLARS Figures 5.8 and 5.9, respectively. They are similarly
read and interpreted as the LIFETIME DOLLARS multivariable CHAID
trees.

5.5 CHAID Market Segmentation

I take this opportunity to use the analysis (so far) to illustrate CHAID as
a market segmentation technique. A closer look at the two full CHAID
trees in Figures 5.4 and 5.5 identifies three market segments pairs, which
show three levels of response performance, 0.76%/0.77%, 4.8%/4.6% and
1.41%/1.46%. Accordingly, CHAID provides the cataloguer with marketing
intelligence for high, medium and low performing segments. Marketing
strategy can be developed to stimulate the high-performers with techniques such as cross-selling, or pique interest in the medium-performers

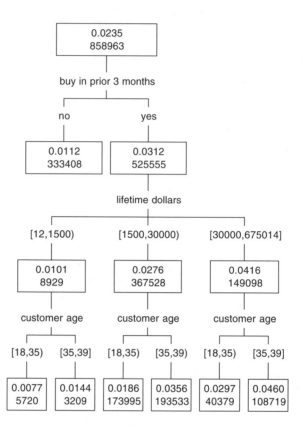

FIGURE 5.9
Multivariable CHAID Tree for Effect of CUST_AGE, accounting for LIFETIME DOLLARS and
PRIOR_BY = yes

with new products; as well as prod the low-performers with incentives
and discounts.

The descriptive profiles of the three market segments are as follows:

> *Market segment #1:* Customers whose ages are in the interval [18, 35)
> and have *not* purchased in prior three months and have lifetime
> dollars in the interval [1500, 30000). AEP of response is 0.0076. See
> branch #3 in Figure 5.4. Customers whose ages are in the interval
> [18, 35) and have purchased in prior three months and have lifetime
> dollars in the interval [12, 1500). AEP of RESPONSE is 0.0077. See
> branch #1 in Figure 5.5.

> *Market segment #2:* Customers whose ages are in the interval [35, 93]
> and have *not* purchased in prior three months and have lifetime
> dollars in the interval [12, 1500). AEP of response is 0.0048. See
> branch #4 in Figure 5.4. Customers whose ages are in the interval
> [35, 93] and have purchased in prior three months and have lifetime

dollars in the interval [30000, 675014]. AEP of RESPONSE is 0.0460. See branch #6 in Figure 5.5.

Market segment #3: Customers whose ages are in the interval [35, 93] and have *not* purchased in prior three months and have lifetime dollars in the interval [1500, 30000). AEP of RESPONSE is 0.0141. See branch #5 in Figure 5.4. Customers whose ages are in the interval [35, 93] and have purchased in prior three months and have lifetime dollars in the interval [12, 1500). AEP of RESPONSE is 0.0144. See branch #4 in Figure 5.5.

5.6 CHAID Tree Graphs

Displaying the multivariable CHAID trees in a *single* graph provides the desired displays of a *complete* assessment of the effects the predictor variables on response. Construction and interpretation of the *CHAID tree graph* for a given predictor is as follows:

1. Collect the set of multivariable CHAID trees for a given predictor variable. For example, for PRIOR_BY there are two trees, which correspond to the two values of PRIOR_BY, yes and no.

2. For each branch, plot the AEP of response values (y-axis) and the minimum values of the end-leaf nodes of the given predictor variable (x-axis).[4]

3. For each branch, connect the *nominal points* (AEP response value, minimum value). The resultant *trace line* segment represents a market or customer segment defined by the branch's intermediate-leaf intervals/nodes.

4. The shape of the trace line indicates the effect of the predictor variable on response for that segment. A comparison of the trace lines provides a total view of how the predictor variable effects response accounting for the presence of the other predictor variables.

The LIFETIME DOLLARS CHAID tree graph in Figure 5.10 is based on the multivariable LIFETIME DOLLARS CHAID trees in Figures 5.4 and 5.5. The top trace line with a noticeable bend corresponds to older customers (age 35 years and up) who have made purchases in the prior three months. The implication is LIFETIME DOLLARS has a nonlinear effect on RESPONSE for this customer segment. As LIFETIME DOLLARS goes from a nominal

---

[4] The minimum value is one of several values which can be used; alternatives are the mean or median of each predefined interval.

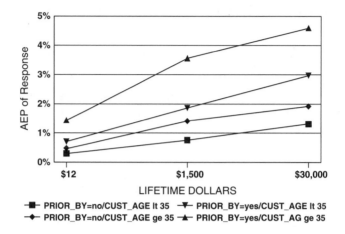

FIGURE 5.10

CHAID Tree Graph for Effects of LIFETIME on REPSONSE by CUST_AGE and PRIOR_BY

$12 to a nominal $1,500 to a nominal $30,000, RESPONSE increases at a nonconstant rate as depicted in the tree graph.

The other trace lines are viewed as straight lines[5] with various slopes. The implication is LIFETIME DOLLARS has various constant effects on RESPONSE across the corresponding customer segments. As LIFETIME DOLLARS goes from a nominal $12 to a nominal $1,500 to a nominal $30,000, RESPONSE increases at various constant rates as depicted in the tree graph.

The PRIOR_BY CHAID tree graph in Figure 5.11 is based on the multivariable PRIOR_BY CHAID trees in Figures 5.6 and 5.7. I focus on the slopes of the trace lines.[6] The evaluation rule is as follows: the steeper the slope, the greater the constant effect on RESPONSE. Among the six trace lines, the top trace line for the "top" customer segment of older customers with lifetime dollars equal to/greater than $30,000 has the steepest slope. The implications are: PRIOR_BY has a rather noticeable constant effect on RESPONSE for the top customer segment; the size of the PRIOR_BY effect for the top customer segment is greater than the PRIOR_BY effect for the remaining five customer segments. As PRIOR_BY goes from 'no' to 'yes,' RESPONSE increases at a rather noticeable constant rate for the top customer segment as depicted in the tree graph.

The remaining five trace lines have slopes of various steepness. The implication is PRIOR_BY has various constant effects on RESPONSE across the corresponding five customer segments. As PRIOR_BY goes from 'no' to 'yes,' RESPONSE increases at various constant rates as depicted in the tree graph.

---

[5] The segment that is defined by PRIOR_BY=no and CUST_AGE greater than/equal 35 years appears to have a very slight bend. However, I treat its trace line as straight because the bend is very slight.

[6] The trace lines are necessarily straight because two points (PRIOR_BY points 'no' and 'yes') always determine a straight line.

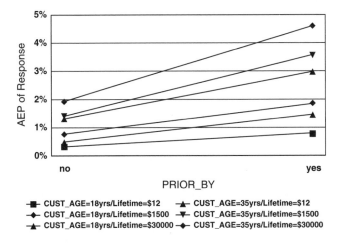

FIGURE 5.11
CHAID Tree Graph for Effects of PRIOR_BY on REPSONSE by CUST_AGE and LIFETIME DOLLARS

The CUST_AGE CHAID tree graph in Figure 5.12 is based on the multi-variable CUST_AGE CHAID trees in Figures 5.8 and 5.9. There are two sets of parallel trace lines with different slopes. The first set of the top two parallel trace lines correspond to two customer segments defined by:

1. PRIOR_BY = no and LIFETIME DOLLARS in the interval [1500, 30000)

2. PRIOR_BY = yes and LIFETIME DOLLARS in the interval [30000, 675014].

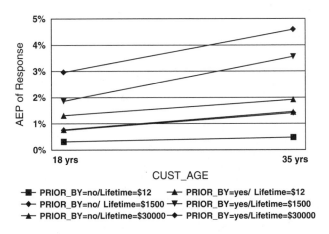

FIGURE 5.12
CHAID Tree Graph for Effects of CUST_AGE on REPSONSE by PRIOR_BY and LIFETIME DOLLARS

The implication is CUST_AGE has the same constant effect on RESPONSE for the two customer segments. As CUST_AGE goes from a nominal age of 18 years to a nominal age of 35 years, RESPONSE increases at a constant rate as depicted in the tree graph.

The second set of the next three parallel trace lines (two trace lines virtually overlap each other) correspond to three customer segments defined by:

1. PRIOR_BY = no and LIFETIME DOLLARS in the interval [30000, 675014]

2. PRIOR_BY = yes and LIFETIME DOLLARS in the interval [1500, 30000)

3. PRIOR_BY = yes and LIFETIME DOLLARS in the interval [12, 1500].

The implication is CUST_AGE has the same constant effect on RESPONSE for the three customer segments. As CUST_AGE goes from a nominal age of 18 years to a nominal age of 35 years, RESPONSE increases at a constant rate as depicted in the tree graph. Note that the constant CUST_AGE effect for the three customer segments is less than the CUST_AGE effect for the former two customer segments, as the slope of the former segments is less steep than that of the latter segements.

Last, the bottom trace line, which corresponds to the customer segment defined by PRIOR_BY = no and LIFETIME DOLLARS in the interval [12, 1500) has virtually no slope, as it is nearly horizontal. The implication is CUST_AGE has no effect on RESPONSE for the corresponding customer segment.

5.7   Summary

After a brief introduction of the logistic regression model as the standard technique for building a response model in database marketing, I focused on its interpretation, an area in the literature that has not received a lot of attention. I discussed the traditional approach to interpreting a logistic regression model using the odds-ratio statistic, which measures the effect of a predictor variable on the odds of response. Then I introduced two weaknesses of the odds ratio. One, the odds ratio is reported in terms of unwieldy odds-of-responding-yes units. Two, the odds ratio only provides an incomplete assessment of a predictor variable's effect on response, since it does not explicitly reflect the relationship of the other predictor variables.

I proposed a CHAID-based method to supplement the odds ratio, as it supplements its two weaknesses. The CHAID-based method adds a visual touch to the original concept of the odds ratio. I illustrated the new method, which exports the information of the odds ratio into CHAID trees, visual displays in terms of simple probability values. More important, the CHAID-based method makes possible the desired complete assessment of a predictor variable's effect on response explicitly reflecting the relationship of the other predictor variables. I illustrated the new method, which combines individual predictor variable CHAID tree into a multivariable CHAID tree graph as a visual complete assessment of a predictor variable.

Moreover, I pointed out that the CHAID-based method can be used to enhance the interpretation of any model — response or profit — built by either statistical or machine learning techniques. The only assumption of the method — that the predicted dependent variable values along with the predictor variables used in the modeling process are available — allows the new method to be applied to any model.

6

The Importance of the Regression Coefficient

Interpretation of the ordinary regression model — the most popular technique for making predictions of a single continuous variable — focuses on the model's coefficients with the aid of three concepts: the statistical p-value, variables "held constant," and the standardized regression coefficient. The purpose of this chapter is to take a closer look at these widely used, yet often misinterpreted concepts. This chapter demonstrates that the statistical p-value as a sole measure for declaring X_i an important predictor variable is sometimes problematic; second, that the concept of variables "held constant" is critical for reliable assessment of how X_i affects the prediction of Y; and last, that the standardized regression coefficient provides the correct ranking of variables in order of predictive importance under special circumstances.

6.1 The Ordinary Regression Model

The ordinary regression model, formally known as the ordinary least squares multiple linear regression model, is the most popular technique for making predictions of a single continuous variable. Its theory is well established and the estimation algorithm is available in all major statistical computer software packages. The model is relatively easy to build, and almost always produces useable results.

Let Y be a continuous dependent variable, e.g., sales, and X_1, X_2, ..., X_i, ..., X_n be the predictor variables. The regression model (prediction equation) is defined in Equation (6.1):

$$Y = b_0 + b_1X_1 + b_2X_2 + ... + b_iX_i + ... + b_nX_n \qquad (6.1)$$

The bs are the regression coefficients,[1] which are estimated by the method of ordinary least squares. Once the coefficients are estimated, an individual's predicted Y value (estimated sales) is calculated by "plugging-in" the values of the predictor variables for that individual in the equation.

---

[1] b_0 is called the intercept, which serves as a mathematical necessity for the regression model. However, b_0 can be considered the coefficient of $X_0=1$.

6.2   Four Questions

Interpretation of the model focuses on the regression coefficient with the aid of three concepts: the statistical p-value, the average change in Y associated with a unit change in X_i when the other Xs^2 are "held constant," and the standardized regression coefficient. The following four questions universally apply to any discussion of the regression coefficient, and are discussed in detail to provide a better understanding of why the regression coefficient is important:

1. Is X_i important for making good predictions? The usual answer is: X_i is an important predictor variable if its associated p-value is less then 5%. This answer is correct for experimental studies, but may not be correct for big data[3] applications.

2. How does X_i affect the prediction of Y? The usual answer: Y experiences an average change of b_i with a unit increase in X_i when the other Xs are held constant. This answer is an "honest" one, which often is not accompanied by a caveat.

3. Which variables in the model have the greatest effect on the prediction of Y, in ranked order? The usual answer: the variable with the largest regression coefficient has the greatest effect; the variable with the next largest regression coefficient has the next greatest effect, and so on. This answer is almost always incorrect.

4. Which variables in the model are the most important predictors, in ranked order? The usual answer: the variable with the largest standardized regression coefficient is the most important variable; the variable with the next largest standardized regression coefficient is the next important variable, and so on. This answer is almost always incorrect.

6.3   Important Predictor Variables

X_i is declared an important predictor variable if it significantly reduces the regression model's prediction error (actual Y – predicted Y). The size of reduction in prediction error due to X_i can be tested for significance with the null hypothesis-signficance testing procedure. Briefly, I have outlined the procedure[4] as follows:

---

[2] The other Xs consist of n-1 variables without Xi, namely, $X_1, X_2, ..., X_{i-1}, X_{i+1}, ..., X_n$.

[3] Big data are defined in Section 1.6 of Chapter 1.

[4] See any good textbook on the null hypothesis-significance testing procedure, such as *Statistical Significance*, Chow, S.L., Sage Publications, 1996.

1. The null hypothesis (NH) and alternative hypothesis (AH) are defined as follows:

   NH: change in mean squared prediction error due to Xi ($cMSE\_X_i$) is equal to zero.

   AH: $cMSE\_X_i$ is not equal to zero.

2. Significance testing for $cMSE\_X_i$ is equivalent to significance testing for the regression coefficient for Xi, b_i. Therefore, NH and AH can be alternatively stated:

   NH: b_i is equal to zero.

   AH: b_i is not equal to zero (b_i is some nonzero value).

3. The working assumptions[5] of the testing procedure are that the sample size is correct and the sample accurately reflects the relevant population. (Well-established procedures for determining the correct sample size for experimental studies already exist.)

4. The decision to reject or fail-to-reject NH is based on the statistical p-value.[6] The statistical p-value is the probability of observing a value of the sample statistic (cMSE or b_i) as extreme or more extreme than the observed value (sample evidence) given NH is true.[7]

5. The decision rule:

   a) If the p-value is *not* "very small," typically greater than the 5%, then the sample evidence supports the decision to fail-to-reject NH.[8] It is concluded that b_i is zero and X_i does not significantly contribute to the reduction in prediction error. Thus, X_i is not an important predictor variable.

   b) If the p-value is very small, typically less than the 5%, then the sample evidence supports the decision to reject NH in favor of accepting AH. It is concluded that b_i (or $cMSE\_X_i$) has some nonzero value and X_i contributes a significant reduction in prediction error. Thus, X_i is an important predictor variable.

The decision rule makes it is clear that the p-value is an indicator of the *likelihood* that the variable has *some* predictive importance, *not* an indicator of how *much* importance (AH does not specify a value of b_i). Thus, *a smaller p-value implies a greater likelihood of some predictive importance, not a greater predictive importance.* This is contrary to the common misinterpretation of the p-value: the smaller the p-value, the greater predictive importance of the associated variable.

---

[5] There is a suite of classical assumptions that are required for the proper testing of the least-squares estimate of b_i. See any good mathematical statistics textbook, such as *Modern Regression Methods*, Ryan, T. P. Wiley, 1997.

[6] Fail to reject NH is not equivalent to accepting NH.

[7] The p-value is a conditional probability.

[8] The choice of "very small" is arbitrary; but convention set it at 5% or less.

6.4 P-Values and Big Data

Relying solely on the p-value for declaring important predictor variables is problematic in big data applications, which are characterized by "large" samples drawn from populations with unknown spread of the Xs. The p-value is affected by the sample size — as sample size increases the p-value decreases, and by the spread of X_i — as spread of X_i increases the p-value decreases.[9] [1] Accordingly, a small p-value may be due to a large sample and/or a large spread of the Xs. Thus, *in big data applications, a small p-value is only an indicator of a potentially important predictor variable.*

The issue of how the p-value is affected by moderating sample size is currently unresolved. Big data are nonexperimental data for which there are no procedures for determining the correct large sample size. A large sample produces many small p-values — a spurious result. The associated variables are often declared important when in fact they are not important, which reduce the stability of a model.[10, 11] [2] A procedure to adjust the p-values when working with big data is needed.[12] Until such procedures are established, the recommended ad hoc approach is as follows: *in big data applications, variables with small p-values must undergo a final assessment of importance based on their actual reduction in prediction error. Variables associated with the greatest reduction in prediction error can be declared important predictors.* When problematic cases are eliminated, the effects of spread of the Xs can be moderated. For example, if the relevant population consists of 35–65 year olds, and the big data includes 18–65 year olds, then simply excluding the 18–34 year olds eliminates the spurious effects of spread.

6.5 Returning to Question #1

Is X_i important for making good predictions? The usual answer: X_i is an important predictor variable if its associated p-value is less then 5%.

This answer is correct for experimental studies, in which sample sizes are correctly determined and samples have presumably known spread. For big data applications, in which large samples with unknown spread adversely affect the p-value, a small p-value is only an indicator of a potentially impor-

---

[9] The size of b_i also effects the p-value: as b_i increases the p-value decreases. This factor cannot be controlled by the analyst.

[10] Falsely reject NH.

[11] Fasely fail-to-reject NH. The effect of a "small" sample is that a variable can be declared unimportant when, in fact, it is important.

[12] A procedure similar to the Bonferroni method, which adjusts p-values downward because repeated testing increases the chance of incorrectly declaring a relationship or coefficient significant.

tant predictor variable. The associated variables must go through an ad hoc evaluation of their actual reduction in the prediction errors before ultimately being declared important predictors.

6.6   Predictor Variable's Effect on Prediction

An assessment of the effect of predictor variable X_i on the prediction of Y focuses on the regression coefficient b_i. The common interpretation is that b_i is the average change in the predicted Y value associated with a unit change in X_i when the other Xs are "held constant." A detailed discussion of what b_i actually measures and how it is calculated shows that this is an "honest" interpretation, which must be accompanied by a caveat.

The regression coefficient b_i, also known as a partial regression coefficient, is a measure of the linear relationship between Y and X_i when the influences of the other Xs are partialled out or "held constant." The expression "held constant" implies that the calculation of b_i involves the removal of the effects of the other Xs. While the details of the calculation are beyond the scope of this chapter, it suffices to outline the steps involved, as delineated below.[13]

The calculation of b_i uses a method of statistical control in a three-step process:

1. The removal of the linear effects of the other Xs from Y produces a new variable Y-adj (= Y-adjusted-linearly-for-the other Xs).
2. The removal of the linear effects of the other Xs from X_i produces a new variable X_i-adj (= X_i-adjusted-linearly-for-the other Xs).
3. The regression of Y-adj on X_i-adj produces the desired partial regression coefficient b_i.

The partial regression coefficient b_i is an "honest" estimate of the relationship between Y and X_i (with the effects of the other Xs partialled out), because it is based on statistical control, not experimental control. The statistical control method estimates b_i *without* data for which the relationship between Y and X_i is actually observed when the other Xs are actually held constant. The rigor of this method ensures the estimate is an "honest" one.

In contrast, the experimental control method involves collecting data, for which the relationship between X_i and Y is *actually* observed when the other Xs are *actually* held constant. The resultant partial regression coefficient is directly measured and therefore yields a "true" estimate of b_i. Regrettably, experimental control data are difficult and expensive to collect.

---

[13] The procedure of statistical control can be found in most basic statistics textbooks.

6.7   The Caveat

There is one caveat to insure the proper interpretation of the partial regression coefficient. *It is not enough to know the variable being multiplied by the regression coefficient; the other Xs must also be known.* [3] Recognition of the other Xs in terms of their values in the sample insures that interpretation is valid. Specifically, the average change b_i in Y is valid for each and every unit change in X_i, within the range of X_i values in the sample when the other Xs are held constant within the ranges or region of the other Xs values in the sample.[14] This point is clarified in the following illustration.

I regress SALES (in dollar units) on EDUCATION (in year units), AGE (in year units), GENDER (in female-gender units; 1 = female and 0 = male), and INCOME (in thousand-dollar units). The regression Equation is defined in Equation (6.2):

$$SALES = 68.5 + 0.75*AGE + 1.03*EDUC + 0.25*INCOME + 6.49*GENDER$$
(6.2)

Interpretation of the regression coefficients is as follows:

1. The individual variable ranges in Table 6.1 suffice to mark the boundary of the region of the other Xs values.
2. For AGE, the average change in SALES is 0.75 dollars for each and every 1-year increase in AGE within the range of AGE values in the sample when EDUC, INCOME and GENDER (E-I-G) are held constant within the E-I-G region in the sample.
3. For EDUC, the average change in SALES is 1.03 dollars for each and every 1-year increase in EDUC within the range of EDUC values in the sample when AGE, INCOME and GENDER (A-I-G) are held constant within the (A-I-G) region in the sample.
4. For INCOME, the average change in SALES is 0.25 dollars for each and every $1000 increase in INCOME within the range of INCOME values in the sample when AGE, EDUC and GENDER (A-E-G) are held constant within the A-E-G region in the sample.
5. For GENDER, the average change in SALES is 6.49 for a "FEMALE-GENDER" unit increase (a change from male to female) when AGE, INCOME and EDUCATION (A-I-E) are held constant within the A-I-E region in the sample.

To further the discussion on the proper interpretation of the regression coefficient I consider a composite variable (e.g., $X_1 + X_2$, X_1*X_2, or X_1/X_2)

---

[14] The region of the other Xs values is defined as the values in the sample common to all the individual other Xs-variable ranges.

TABLE 6.1

Descriptive Statistics of Sample

Variable	Mean	Ranges (min, max)	StdDev	H-spread
SALES	30.1	(8, 110)	23.5	22
AGE	55.8	(44, 76)	7.2	8
EDUC	11.3	(7, 15)	1.9	2
INCOME	46.3	(35.5, 334.4)	56.3	28
GENDER	0.58	(0,1)	0.5	1

which often is included in regression models. I show that the regression coefficients of the composite variable and the variables defining the composite variable are not interpretable.

I add a composite product variable to the original regression model: EDUC_INC (= EDUC*INCOME, in years*thousand-dollar units). The resultant regression model is in Equation (6.3):

$$SALES = 72.3 + 0.77*AGE + 1.25*EDUC + 0.17*INCOME +$$
$$6.24*GENDER + 0.006*EDUC\_INC \qquad (6.3)$$

Interpretation of the new regression model and its coefficients is as follows:

1. The coefficients of the original variables have changed. This is expected because the value of the regression coefficient for X_i depends not only on the relationship between Y and X_i, but on the relationships between the other Xs and X_i, and the other Xs and Y.

2. Coefficients for AGE and GENDER changed from 0.75 to 0.77, and 6.49 to 6.24, respectively.

3. For AGE, the average change in SALES is 0.77 dollars for each and every 1-year increase in AGE within the range of AGE values in the sample when EDUC, INCOME, GENDER and EDUC_INC (E-I-G-E_I) are held constant within the E-I-G-E_I region in the sample.

4. For GENDER, the average change in SALES is 6.24 dollars for a "FEMALE-GENDER" unit increase (a change from male to female) when AGE, EDUC, INCOME, and EDUC_INC (A-E-I-E_I) are held constant within the A-E-I-E_I region in the sample.

5. Unfortunately, the inclusion of EDUC_INC in the model compromises the interpretation of the regression coefficients for EDUC and INCOME — two variables that cannot be interpreted! Consider the following:

   a) For EDUC, the usual interpretation is that the average change in SALES is 1.25 dollars for each and every 1-year increase in EDUC within the range of EDUC values in the sample when AGE, INCOME, GENDER and EDUC_INC (A_I-G-E_I) are held constant within the A-I-G-E_I region in the sample. *This statement is meaningless.* It is not possible to hold constant EDUC_INC for any 1-year increase in EDUC: as EDUC would vary, so must EDUC_INC. Thus, no meaningful interpretation can be given to the regression coefficient for EDUC.

b) Similarly, no meaningful interpretations can be given to the regression coefficients for INCOME and EDUC_INC. It is not possible to hold constant EDUC_INC for INCOME. Moreover, for EDUC_INC, it is not possible to hold constant EDUC and INCOME.

6.8   Returning to Question #2

How does the X_i affect the prediction of Y? The usual answer: Y experiences an average change of b_i with a unit increase in X_i when the other Xs are held constant.

This answer is an "honest" one (because of the statistical control method that estimates b_i), which must be accompanied by the values of the X_i-range and the other Xs-region. Unfortunately, the effects of a composite variable and the variables defining the composite variable cannot be assessed because their regression coefficients are not interpretable.

6.9   Ranking Predictor Variables by Effect On Prediction

I return to the first regression model in Equation (6.2).

$$\text{SALES} = 68.5 + 0.75*\text{AGE} + 1.03*\text{EDUC} + 0.25*\text{INCOME} + 6.49*\text{GENDER}$$
(6.2)

A common misinterpretation of the regression coefficient is that GENDER has the greatest effect on SALES, followed by EDUC, AGE and INCOME, because the coefficients can be ranked in that order. The problem with this interpretation is discussed below. I have also presented the correct rule for ranking predictor variables in terms of their effects on the prediction of dependent variable Y.

This regression model illustrates the difficulty in relying on the regression coefficient for ranking predictor variables. The regression coefficients are incomparable because different units are involved. No meaningful comparison can be made between AGE and INCOME, because the variables have different units (years and thousand-dollars units, respectively). Comparing GENDER and EDUC is another mixed-unit comparison between female-gender and years, respectively. Even a comparison between AGE and EDUC whose units are the same (i.e., years) is problematic, because the variables have unequal spreads (e.g., standard deviation, StdDev), in Table 6.1.

The correct ranking of predictor variables in terms of their effects on the prediction of Y is the ranking of the variables by the magnitude of the standardized regression coefficient. The sign of the standardized coefficient is disregarded, as it only indicates direction. (Exception to this rule will be discussed later.) The standardized regression coefficient (also known as the beta regression coefficient) is produced by multiplying the original regression coefficient (also called raw regression coefficient) by a conversion factor (CF). The standardized regression coefficient is unit-less, just a plain number allowing meaningful comparisons among the variables. The transformation equation that converts a unit-specific raw regression coefficient into a unit-less standardized regression coefficient is in Equation (6.4):

Standardized reg. coefficient for Xi = CF * Raw reg. coefficient for Xi (6.4)

The conversion factor is defined as the ratio of a unit measure of Y-variation to a unit measure of Xi-variation. The standard deviation (StdDev) is the usual measure used. However, if the distribution of variable is not bell-shaped, then the standard deviation is not reliable, and the resultant standardized regression coefficient is questionable. An alternative measure, one that is not affected by the shape of the variable, is the H-spread. The H-spread is defined as the difference between the 75-percentile and 25-percentile of the variables distribution. Thus, there are two popular CF's, and two corresponding transformations in Equations (6.5) and (6.6).

Standardized reg., coefficient for Xi =
[StdDev of Xi/StdDev of Y] * Raw reg., coefficient for Xi   (6.5)

Standardized reg., coefficient for Xi =
[H-spread of Xi/H-spread of Y] * Raw reg., coefficient for Xi   (6.6)

Returning to the illustration of the first regression model, I note the descriptive statistics in Table 6.1. AGE has equivalent values for both StdDev and H-spread (23.5 and 22); so does EDUC (7.2, and 8). INCOME, which is typically not bell-shaped, has quite different values for the two measures, the unreliable StdDev is 56.3 and the reliable H-spread is 28.

Dummy variables have no meaningful measure of variation. StdDev and H-spread are often reported for dummy variables as a matter of course, but they have "no value" at all.

TABLE 6.2

Raw and Standardized Regression Coefficients

Variable (unit)	Raw Coefficients (unit)	Standardized Coefficient (unit-less) based on	
		StdDev	H-spread
AGE (years)	0.75 (dollars/years)	0.23	0.26
EDUC (years)	1.03 (dollars/years)	0.09	0.09
INCOME			
(000-dollars)	0.25 (dollars/000-dollars)	0.59	0.30
GENDER			
(female-gender)	6.49 (dollars/female-gender)	0.14	0.27

The correct ranking of the predictor variables in terms of their effects on SALES — based on the magnitude of the standardized regression coefficients in Table 6.2 — puts INCOME first, with the greatest effect, followed by AGE and EDUC. This ordering is obtained with either the StdDev or the H-spread. However, INCOME's standardized coefficient should be based on the H-spread, as INCOME is typically skewed. Because GENDER is a dummy variable with no meaningful CF, its effect on prediction of Y cannot be ranked.

6.10 Returning to Question #3

Which variables in the model have the greatest effect on the prediction of Y, in ranked order? The usual answer: the variable with the largest regression coefficient has the greatest effect; the variable with the next largest regression coefficient has the next greatest effect, and so on. This answer is almost always incorrect.

The correct ranking of predictor variables in terms of their effects on the prediction of Y is the ranking of the variables by the magnitude of the standardized regression coefficient. Predictor variables that cannot be ranked are (1) dummy variables, which have no meaningful measure of variation; and (2) composite variables and the variables defining them, which have both raw regression coefficients and standardized regression coefficients that cannot be interpreted.

6.11 Returning to Question #4

Which variables in the model are the most important predictors, in ranked order? The usual answer: the variable with the largest standardized regression coefficient is the most important variable; the variable with the next largest standardized regression coefficient is the next important variable, and so on.

This answer is correct *only* when the predictor variables are *uncorrelated*. With uncorrelated predictor variables in a regression model, there is a rank-order correspondence between the magnitude of the standardized coefficient and reduction in prediction error. Thus, the magnitude of the standardized coefficient can rank the predictor variables in order of most-to-least importance. Unfortunately, there is *no* rank-order correspondence with correlated predictor variables. Thus, the magnitude of standardized coefficient of correlated predictor variables *cannot* rank the variables in terms of predictive importance. The proof of these facts is beyond the scope of this chapter. [4]

6.12 Summary

Thus, it should now be clear that the common misconceptions of the regression coefficient lead to an incorrect interpretation of the ordinary regression model. This exposition puts to rest these misconceptions, promoting an appropriate and useful presentation of the regression model.

Common misinterpretations of the regression coefficient are problematic. The use of the statistical p-value as a sole measure for declaring X_i an important predictor is problematic because of its sensitivity to sample size and spread of the X values. In experimental studies, the analyst must insure that the study design takes into account these sensitivities to allow for valid inferences to be drawn. In big data applications, variables with small p-values must undergo a final assessment of importance based on their actual reduction in prediction error. Variables associated with the greatest reduction in prediction error can be declared important predictors.

When assessing how X_i affects the prediction of Y, the analyst must report the other Xs in terms of their values. Moreover, the analyst must not attempt to assess the effects of a composite variable and the variables defining the composite variable because their regression coefficients are not interpretable.

By identifying the variables in the model that have the greatest effect on the prediction of Y — in rank order — the superiority of the standardized regression coefficient for providing correct ranking of predictor variables, rather than the raw regression coefficient, should be apparent. Furthermore, it is important to recognize that the standardized regression coefficient ranks variables in order of predictive importance (most to least) for only uncorrelated predictor variables. This is not true for correlated predictor variables. It is imperative that the analyst does not use the coefficient to rank correlated predictor variables despite the allure of its popular misuse.

References

1. Kraemer, H.C. and Thiemann, S. *How Many Subjects?* Sage Publications, Thousand Oaks, CA, 1987.
2. Dash, M. and Liu, H., Feature Selection for Classification, *Intelligent Data Analysis*, Elsevier Science, New York, 1997.
3. Mosteller, F. and Tukey, J., *Data Analysis and Regression*, Addison-Wesley, Reading, MA, 1977.
4. Hayes, W.L., *Statistics for the Social Sciences*, Holt, Rinehart and Winston, Austin, TX, 1972.

7

The Predictive Contribution Coefficient: A Measure of Predictive Importance

Determining which variables in a model are its most important predictors (in ranked order) is a vital element in the interpretation of a linear regression model. A general rule is to view the predictor variable with the largest standardized regression coefficient as the most important variable; the predictor variable with the next largest standardized regression coefficient as the next important variable, and so on. This rule is intuitive, easy to apply and provides practical information for understanding how the model works. Unknown to many, however, is that the rule is theoretically problematic. Thus, the purpose of this chapter is twofold: first, to discuss why the decision rule is theoretically amiss, yet works well in practice; second, to present an alternative measure — the *predictive contribution coefficient* — which offers greater utile information than the standardized coefficient, as it is an assumption-free measure founded in the data mining paradigm.

7.1 Background

Let Y be the dependent variable, and X_1, X_2, ..., X_i, ..., X_n be the predictor variables. The linear regression model is defined in Equation (7.1):

$$Y = b_0 + b_1{}^*X_1 + b_2{}^*X_2 + ... + b_i{}^*X_i + ... + b_n{}^*X_n \qquad (7.1)$$

The bs are the raw regression coefficients, which are estimated by the methods of ordinary least-squares and maximum likelihood for a continuous and binary-dependent variable, respectively. Once the coefficients are estimated, an individual's predicted Y value is calculated by 'plugging-in' the values of the predictor variables for that individual in the equation.

The interpretation of the raw regression coefficient points to the following question: how does X_i affect the prediction of Y? The answer is that the predicted Y experiences an average change of b_i with a unit increase in X_i

when the other Xs are held constant. A common misinterpretation of the raw regression coefficient is that predictor variable with the largest value (the sign of coefficient is ignored) has the greatest effect on the predicted Y. Unless the predictor variables are measured in the same units, the raw regression coefficient values can be so unequal that they cannot be compared with each other. The raw regression coefficients must be standardized to import the different units of measurement of the predictor variables to allow a fair comparison.

The standardized regression coefficients are just plain numbers, like "points," allowing material comparisons among the predictor variables. The standardized regression coefficient (SRC) can be determined by multiplying the raw regression coefficient by a conversion factor, a ratio of a unit measure of X_i-variation to a unit measure of Y-variation. The SRC in an ordinary regression model is defined in Equation (7.2), where StdDevX$_i$ and StdDevY are the standard deviations for X_i and Y, respectively.

$$\text{SRC for } X_i = (\text{StdDevX}_i/\text{StdDevY}) * \text{Raw reg., coefficient for } X_i \quad (7.2)$$

For the logistic regression model, the problem of calculating the standard deviation of the dependent variable, which is the logit Y, not Y, is complicated. The problem has received attention in the literature with solutions that provide inconsistent results. [1] The simplest is the one used in the SAS system, although it is not without its problems. The StdDev for the logit Y is taken as 1.8138 (the value of the standard deviation of the standard logistic distribution). Thus, the SRC in logistic regression model is defined in Equation (7.3).

$$\text{SRC for } X_i = (\text{StdDevXi}/1.8138) * \text{Raw reg., coefficient for } X_i \quad (7.3)$$

The SRC can also be obtained directly by performing the regression analysis on standardized data. You will recall that standardizing the dependent variable Y and the predictor variables X_is creates new variables zY and zX$_i$, such that their means and standard deviations are equal to zero and one, respectively. The coefficients obtained by regressing zY on the zX$_i$s are, by definition, the standardized regression coefficients.

The question "which variables in the model are its most important predictors, in ranked order?" needs to be annotated before being answered. The importance or ranking is traditionally taken in terms of the statistical characteristic of reduction in prediction error. The usual answer is provided by the decision rule: the variable with the largest SRC is the most important variable; the variable with the next largest SRC is the next important variable, and so on. This decision rule is correct with the *unnoted caveat that the predictor variables are uncorrelated*. There is a rank-order correspondence between the SRC (the sign of coefficient is ignored) and the reduction in prediction error in a regression model with only uncorrelated predictor variables. There is one other unnoted caveat for the proper use of the decision rule: the SRC for a dummy predictor variable (defined by only two values) is not reliable, as the standard deviation of a dummy variable is not meaningful.

Database regression models in virtually every case have correlated predictor variables, which challenge the utility of the decision rule. Yet, the rule continuously provides useful information for understanding how the model works without raising sophistic findings. The reason for its found utility is that there is an unknown working assumption at play and met in database applications: the reliability of the ranking based on the SRC increases as the average correlation among the predictor variables decreases. Thus, for well-built models, which necessarily have a minimal correlation among the predictor variables, the decision rule remains viable in database regression applications. However, there are caveats: dummy variables cannot be reliably ranked; and composite variables (derived from other predictor variables) and the elemental variables (defining the composite variables) are inherently highly correlated and thus cannot be reliably ranked.

7.2 Illustration of Decision Rule

Consider RESPONSE (0 = no; 1 = yes), PROFIT (in dollars), AGE (in years), GENDER (1 = female, 0 = male) and INCOME (in thousand dollars) for ten individuals in the small data in Table 7.1. I standardized the data producing the standardized variables with the notation used above: zRESPONSE, zPROFIT, zAGE, zGENDER and zINCOME (data not shown).

I perform two ordinary regression analyses based on both the raw data and the standardized data. Specifically, I regress PROFIT on INCOME, AGE, and GENDER, and regress zPROFIT on zINCOME, zAGE, and zGENDER. The raw and standardized regression coefficients based on the *raw data* are

TABLE 7.1

Small Data

ID #	Response	Profit	Age	Gender	Income
1	1	185	65	0	165
2	1	174	56	0	167
3	1	154	57	0	115
4	0	155	48	0	115
5	0	150	49	0	110
6	0	119	40	0	99
7	0	117	41	1	96
8	0	112	32	1	105
9	0	107	33	1	100
10	0	110	37	1	95I

TABLE 7.2

PROFIT Regression Output Based on Raw Small Data

Variable	DF	Parameter Estimate	Standard Error	t Value	Pr > |t|	Standardized Estimate
INTERCEPT	1	37.4870	16.4616	2.28	0.0630	0.0000
INCOME	1	0.3743	0.1357	2.76	0.0329	0.3516
AGE	1	1.3444	0.4376	3.07	0.0219	0.5181
GENDER	1	−11.1060	6.7221	−1.65	0.1496	−0.1998

in the Parameter Estimate and Standardized Estimate columns, in Table 7.2, respectively. The raw regression coefficients for INCOME, AGE, and GENDER are 0.3743, 1.3444 and −11.1060, respectively. The standardized regression coefficients for INCOME, AGE, and GENDER are 0.3516, 0.5181 and −0.1998, respectively.

The raw and standardized regression coefficients based on the *standardized data* are in the Parameter Estimate and Standardized Estimate columns, in Table 7.3, respectively. As expected, the raw regression coefficients — which are now the standardized regression coefficients — *equal* the values in the Standardized Estimate column: for zINCOME, zAGE, and zGENDER, the values are 0.3516, 0.5181 and −0.1998, respectively.

I calculate the average correlation among the three predictor variables, which is a moderate 0.71. Thus, the SRC can arguably provide a reliable ranking of the predictor variables for PROFIT. AGE is the most important predictor variable, followed by INCOME; GENDER's ranked position is undetermined.

I perform two logistic regression analyses based on both the raw data and the standardized data. Specifically, I regress RESPONSE on INCOME, AGE, and GENDER, and regress RESPONSE[1] on zINCOME, zAGE, and zGENDER. The raw and standardized logistic regression coefficients based on the *raw data* are in the Parameter Estimate and Standardized Estimate columns, in Table 7.4, respectively. The raw logistic regression coefficients for INCOME, AGE, and GENDER are 0.0680, 1.7336 and 14.3294, respectively. The standardized logistic regression coefficients for zINCOME, zAGE, and zGENDER are 1.8339, 19.1800 and 7.3997, respectively.

TABLE 7.3

PROFIT Regression Ouput Based on Standardized Small Data

Variable	DF	Parameter Estimate	Standard Error	t Value	Pr > |t|	Standardized Estimate
INTERCEPT	1	−4.76E-16	0.0704	0.0000	1.0000	0.0000
zINCOME	1	0.3516	0.1275	2.7600	0.0329	0.3516
zAGE	1	0.5181	0.1687	3.0700	0.0219	0.5181
zGENDER	1	−0.1998	0.1209	−1.6500	0.1496	−0.1998

---

[1] I choose not to use zRESPONSE. Why?

TABLE 7.4

RESPONSE Regression Output Based on Raw Small Data

Variable	DF	Parameter Estimate	Standard Error	Wald Chi-Square	Pr > ChiSq	Standardized Estimate
INTERCEPT	1	−99.5240	308.0000	0.1044	0.7466	
INCOME	1	0.0680	1.7332	0.0015	0.9687	1.0111
AGE	1	1.7336	5.9286	0.0855	0.7700	10.5745
GENDER	1	14.3294	82.4640	0.0302	0.8620	4.0797

TABLE 7.5

RESPONSE Regression Output Based on Standardized Small Data

Variable	DF	Parameter Estimate	Standard Error	Wald Chi-Square	Pr > ChiSq	Standardized Estimate
INTERCEPT	1	−6.4539	31.8018	0.0412	0.8392	
zINCOME	1	1.8339	46.7291	0.0015	0.9687	1.0111
zAGE	1	19.1800	65.5911	0.0855	0.7700	10.5745
zGENDER	1	7.3997	42.5842	0.0302	0.8620	4.0797

The raw and standardized logistic regression coefficients based on the *standardized data* are in the Parameter Estimate and Standardized Estimate columns, in Table 7.5, respectively. Unexpectedly, the raw logistic regression coefficients — which are now the standardized logistic regression coefficients — *do not equal* the values in the Standardized Estimate column. If the ranking of predictor variables is required of the standardized coefficient, this inconsistency presents no problem, as the raw and standardized values produce the same rank order. If information about the expected increase in the predicted Y is required, I prefer the standardized regression coefficients in the Parameter Estimate column, as it follows the definition of SRC.

As determined previously, the average correlation among the three-predictor variables is a moderate 0.71. Thus, the SRC can arguably provide a dependable ranking of the predictor variables for RESPONSE. AGE is the most important predictor variable, followed by INCOME; GENDER's ranked position is undetermined.

7.3   Predictive Contribution Coefficient

The predictive contribution coefficient (PCC) is a development in the data-mining paradigm. It has the hallmarks of the data mining paradigm: flexibility — as it is an assumption-free measure that works equally well with ordinary and logistic regression models; practicality and innovation — as it offers greater utile information than the SRC; and, above all, simplicity — as it is easy to understand and calculate, as is evident from the discussion below.

Consider the linear regression model built on standardized data, and defined in Equation (7.4):

$$zY = b_0 + b_1{}^*zX_1 + b_2{}^*zX_2 + ... + b_i{}^*zX_i + ... + b_n{}^*zX_n \qquad (7.4)$$

The predictive contribution coefficient for zX_i, $PCC(zX_i)$, is a measure of zX_is contribution relative to the other variables' contribution to the model's predictive scores. $PCC(zX_i)$ is defined as the average absolute ratio of the zX_i score-point contribution ($zX_i{}^*b_i$) to the other variables' score-point contribution (total predictive score minus zX_is score-point). Briefly, the PCC is read as follows: the larger the $PCC(zX_i)$ value, the more significant part zX_i has in the model's predictions; ergo, the more important zX_i is as a predictor variable. Exactly how the PCC works and what are its benefits over the SRC are discussed in the next section. Now, I provide justification for the PCC.

Justification of the trustworthiness of the PCC is required, as it depends on the SRC, which itself, as discussed above, is not a perfect measure to rank predictor variables. The effects of any "impurities" (biases) carried by the SRC on the PCC are presumably negligible due to the "wash-cycle" calculations of the PCC. The six-step cycle, which is described in the next section, crunches the actual values of the SRC such that any original bias-effects are washed-out.

I'd like now to revisit the question to determine which variables are the most important predictors. The predictive-contribution decision rule is that the variable with the largest PCC is the most important variable; the variable with the next largest PCC is the next important variable, and so on. In other words, the predictor variables can be ranked from most-to-least important based on the descending order of the PCC. Unlike the reduction-in-prediction-error decision rule, there are no presumable caveats for the predictive-contribution decision rule. Correlated predictor variables, including composite and dummy variables, can be thus ranked.

7.4   Calculation of Predictive Contribution Coefficient

Consider the logistic regression model based on the standardized data in Table 7.5. I illustrate in detail the calculation of PCC(zAGE) with the Necessary Data in Table 7.6.

1. Calculate the TOTAL PREDICTED (logit) score for each individual in the data. For individual ID#1, the standardized predictor variable's values, in Table 7.6, which are multiplied by the corresponding standardized logistic regression coefficients, in Table 7.5, produce the total predicted score of 24.3854.

TABLE 7.6

Necessary Data

ID #	zAGE	zINCOME	zGENDER	Total Predicted Score	zAGE Score-Point Contribution	OTHERVARS Score-Point Contribution	zAGE_OTHVARS
1	1.7354	1.7915	-0.7746	24.3854	33.2858	-8.9004	3.7398
2	0.9220	1.8657	-0.7746	8.9187	17.6831	-8.7643	2.0176
3	1.0123	-0.0631	-0.7746	7.1154	19.4167	-12.3013	1.5784
4	0.1989	-0.0631	-0.7746	-8.4874	3.8140	-12.3013	0.3100
5	0.2892	-0.2485	-0.7746	-7.0938	5.5476	-12.6414	0.4388
6	-0.5243	-0.6565	-0.7746	-23.4447	-10.0551	-13.3897	0.7510
7	-0.4339	-0.7678	1.1619	-7.5857	-8.3214	0.7357	11.3105
8	-1.2474	-0.4340	1.1619	-22.5762	-23.9241	1.3479	17.7492
9	-1.1570	-0.6194	1.1619	-21.1827	-22.1905	1.0078	22.0187
10	-0.7954	-0.8049	1.1619	-14.5883	-15.2560	0.6677	22.8483

2. Calculate the zAGE SCORE-POINT contribution for each individual in the data. For individual ID #1, the zAGE score-point contribution is 33.2858 (= 1.7354*19.1800).

3. Calculate the OTHER VARIABLES SCORE-POINT contribution for each individual in the data. For individual ID #1, the OTHER VARI-ABLES score-point contribution is –8.9004 (= 24.3854 – 33.2858).

4. Calculate the zAGE_OTHVARS for each individual in the data. zAGE_OTHVARS is defined as the absolute ratio of zAGE SCORE_POINT contribution to the OTHER VARIABLES SCORE-POINT contribution. For ID #1, zAGE_OTHVARS is 3.7398 (= absolute value of 33.2858/–8.9004).

5. Calculate PCC(zAGE), the average (median) of the zAGE_OTHVARS values: 2.8787. The zAGE_OTHVARS distribution is typically skewed, which suggests that the median is more appropriate than the mean for the average.

I summarize the results the PCC calculations after applying the five-step process for zINCOME and zGENDER (not shown).

1. AGE is ranked the most important predictor variable, next is GEN-DER, and last is INCOME. Their PCC values are 2.8787, 0.3810 and 0.0627, respectively.

2. AGE is the most important predictor variable with the largest and "large" PCC value of 2.8787, which indicates that zAGE's predictive contribution is 2.887 times the combined predictive contribution of zINCOME and zGENDER. The implication is that AGE is clearly driving the model's predictions. In situations where a predictor variable has a large PCC value, it is known as a *key-driver* of the model.

Note that I do not use the standardized variable names (e.g., zAGE instead of AGE) in the summary of findings. This is to reflect that the issue at hand in determining the importance of predictor variables is the content of variable, which is clearly conveyed by the original name, not the technical name, which reflects a mathematical necessity of the calculation process.

7.5 Extra Illustration of Predictive Contribution Coefficient

This section assumes an understanding of the decile analysis table, which is discussed in full detail in Chapter 12. Readers who are not familiar with the decile analysis table may still be able to glean the key points of the following discussion before reading Chapter 12.

The PCC offers greater utile information than the SRC, notwithstanding the SRC's working assumption and caveats. Both coefficients provide an overall ranking of the predictor variables in a model. However, the PCC can extend beyond an overall ranking by providing a ranking at various levels of model performance. Moreover, the PCC allows for the identification of key-drivers — salient features — which the SRC metric cannot legitimately yield. By way of continuing with the illustration, I discuss these two benefits of the PCC.

Consider the decile performance of the logistic regression model based on the standardized data in Table 7.5. The decile analysis in Table 7.7 indicates the model works well, as it identifies the lowly three responders in the top three deciles.

The PCC as presented so far provides an overall-model ranking of the predictor variables. In contrast, the admissible calculations of the PCC at the decile level provides a decile ranking of the predictor variables with a response modulation, ranging from most-to-least likely to respond. To affect a decile-based calculation of the PCC, I rewrite step 5 in Section 7.4, *Calculation of Prediction Contribution Coefficient*:

Step 5: Calculate PCC(zAGE) – the median of the zAGE_OTHVARS values for each decile.

The PCC decile-based Small Data calculations, which are obvious and trivial, are presented to make the PCC concept and procedure clear, and generate interest in its application. Each of the ten individuals is itself a decile, in which the median value is the individual's value. However, this point is also instructional. The reliability of the PCC value is sample-size dependent. In real applications, in which the decile sizes are large to insure the reliability of the median, the PCC-decile analysis is quite informational as to how the predictor variable ranking interacts across the response modulation produced by the model. The decile-PCCs for the RESPONSE model in Table 7.8 are clearly daunting. I present two approaches to analyze and draw implications from the seemingly scattered arrays of PCCs left by the decile-based calculations. The first approach ranks the predictor variables by the decile-PCCs for each decile. The rank values, which are descending from 1 to 3, from most to least important predictor variable, respectively, are in Table 7.9. Next, the decile-PCC rankings are compared with the "Overall" PCC ranking.

The analysis and implications of Table 7.9 are as follows:

1. The "Overall" PCC importance ranking is AGE, GENDER and INCOME, descendingly. The implication is that an inclusive marketing strategy is defined by primary focus on AGE, secondary emphasis on GENDER, and incidentally calling attention to INCOME.

TABLE 7.7

Decile Analysis for RESPONSE Logistic Regression

Decile	Number of Individuals	Number of Responses	Response Rate (%)	Cum Response Rate (%)	Cum Lift
Top	1	1	100	100.0	333
2	1	1	100	100.0	333
3	1	1	100	100.0	333
4	1	0	0	75.0	250
5	1	0	0	60.0	200
6	1	0	0	50.0	167
7	1	0	0	42.9	143
8	1	0	0	37.5	125
9	1	0	0	33.3	111
Bottom	1	0	0	30.0	100
Total	10	3			

TABLE 7.8

Decile-PCC: Actual Values

Decile	PCC(zAGE)	PCC(zGENDER)	PCC(zINCOME)
Top	3.7398	0.1903	0.1557
2	2.0176	0.3912	0.6224
3	1.5784	0.4462	0.0160
4	0.4388	4.2082	0.0687
5	11.3105	0.5313	0.2279
6	0.3100	2.0801	0.0138
7	22.8483	0.3708	0.1126
8	22.0187	0.2887	0.0567
9	17.7492	0.2758	0.0365
Bottom	0.7510	0.3236	0.0541
Overall	2.8787	0.3810	0.0627

2. The decile-PCC importance rankings are in agreement with the overall PCC importance ranking for all but deciles 2, 4 and 6.

3. For decile 2, AGE remains most important, whereas GENDER and INCOME are reversed in their importance, relative to the overall PCC ranking.

4. For deciles 4 and 6, INCOME remains the least important, whereas AGE and GENDER are reversed in their importance, relative to the overall PCC ranking.

5. The implication is that two decile-tactics are called for, beyond the inclusive marketing strategy. For individuals in decile 2, careful planning includes particular prominence given to AGE, secondarily mentions INCOME, and incidentally addresses GENDER. For individuals in deciles 4 and 6, careful planning includes particular prominence given to GENDER, secondarily mentions AGE, and incidentally addresses INCOME.

TABLE 7.9

Decile-PCC: Rank Values

Decile	PCC(zAGE)	PCC(zGENDER)	PCC(zINCOME)
Top	1	2	3
2	1	3	2
3	1	2	3
4	2	1	3
5	1	2	3
6	2	1	3
7	1	2	3
8	1	2	3
9	1	2	3
Bottom	1	2	3
Overall	1	2	3

The second approach determines decile-specific key-drivers by focusing on the actual values of the obsequious decile-PCCs in Table 7.8. In order to put a methodology in place, a measured value of "large proportion of the combined predictive contribution" is needed. Remember that AGE is informally declared a key-driver of the model because of its large PCC value, which indicates that zAGE's predictive contribution represents a large proportion of the combined predictive contribution of zINCOME and zGENDER. Accordingly, I define predictor variable X_i as a key-driver as follows: X_i *is a key-driver if* $PCC(X_i)$ *is greater than* $1/(k-1)$, *where k is the number of other variables in the model; otherwise,* X_i *is not a key-driver.* The value $1/(k-1)$ is, of course, user-defined; but I presuppose that if a single predictor variable's score-point contribution is greater than the rough average score-point contribution of the other variables, it can safely be declared a key driver of model predictions.

The key-driver definition is used to recode the actual PCCs values into 0–1 values, which represent non-key-driver/key-driver. The result is a key-driver table, which serves as a means of formally declaring the decile-specific key-drivers and overall-model key-drivers. In particular, the table reveals key-drivers patterns across the decile relative to the overall-model key-drivers. I use the key-driver definition to recode the actual PCC values in Table 7.9 into key-drivers in Table 7.10.

The analysis and implications of Table 7.10 are as follows:

1. There is a single overall key-driver of the model, AGE.

2. AGE is also the sole key-driver for deciles top, 3, 7, 8, 9 and bottom.

3. Decile 2 is driven by AGE and INCOME.

4. Deciles 4 and 6 are driven by only GENDER.

5. Decile 5 is driven by AGE and GENDER.

6. The implication is that the salient feature of an inclusive marketing strategy is AGE. To insure the effectiveness of the marketing strategy,

TABLE 7.10

Decile-PCC: Key-Drivers

Decile	Age	Gender	Income
Top	1	0	0
2	1	0	1
3	1	0	0
4	0	1	0
5	1	1	0
6	0	1	0
7	1	0	0
8	1	0	0
9	1	0	0
Bottom	1	0	0
Overall	1	0	0

the "AGE" message must be tactically adjusted to fit the individuals in deciles 1 through 6 (the typical range for model implementation). Specifically, for decile 2, the marketing strategist must add INCOME to the AGE message; for deciles 4 and 6, the marketing strategist must center attention to GENDER with an undertone of AGE; and for decile 5, the marketing strategist must add a GENDER component to the AGE message.

7.6 Summary

I briefly reviewed the traditional approach of using the magnitude of the raw or standardized regression coefficients in determining which variables in a model are its most important predictors. I explained why neither coefficient yields a perfect "importance" ranking of predictor variables. The raw regression coefficient does not account for inequality in the units of the variables. Furthermore, the standardized regression coefficient is theoretically problematic when the variables are correlated, as is virtually the situation in database applications. With a small data set, I illustrated for ordinary and logistic regression models the often-misused raw regression coefficient, and the dubious use of the standardized regression coefficient in determining the rank importance of the predictor variables.

I pointed out that the ranking based on standardized regression coefficient provides useful information without raising sophistic findings. An unknown working assumption is that the reliability of the ranking based on the standardized regression coefficient increases as the average correlation among the predictor variables decreases. Thus, for well-built models, which necessarily have a minimal correlation among the predictor variables, the resultant ranking is accepted practice.

Then, I presented the predictive contribution coefficient, which offers greater utile information than the standardized coefficient. I illustrated how it works, as well as its benefits over the standardized regression coefficient using the small data set.

Last, I provided an extra illustration of the benefits of the new coefficient over the standardized regression coefficient. First, it can rank predictor variables at a decile-level of model performance. Second, it can identify a newly defined predictor variable type, namely, the key-driver. The new coefficient allows the data analyst to determine at the overall model-level and the individual decile-levels which predictor variables are the predominant or key-driver variables of the model's predictions.

Reference

1. Menard, S. *Applied Logistic Regression Analysis*, Sage Publications. Series: Quantitative Applications in the Social Sciences, Thousand Oaks, CA, 1995.

8

CHAID for Specifying a Model with Interaction Variables

In order to increase a model's predictive power beyond that provided by its components, data analysts create an interaction variable, which is the product of two or more component variables. However, a compelling case can be made for utilizing CHAID as an alternative method for specifying a model, thereby justifying the omission of the component variables under certain circumstances. Database marketing provides an excellent example for this alternative method. I illustrate the alternative method with a response model case study.

8.1 Interaction Variables

Consider variables X_1 and X_2. The product of these variables, denoted by X_1X_2, is called a *two-way* or *first-order interaction variable*. An obvious property of this interaction variable is that its information or variance is shared with both X_1 and X_2. In other words, X_1X_2 has inherent high correlation with both X_1 and X_2.

If a third variable (X_3) is introduced, then the product of the three variables ($X_1X_2X_3$) is called a three-way, or second-order interaction variable. It is also highly correlated with each of its component variables. Simply multiplying the component variables can create higher order variables. However, interaction variables of an order greater than three are rarely justified by theory or empirical evidence.

When data have highly correlated variables, it has a condition known as multi-collinearity. When high correlation exists because of the relationship among these variables, the multi-collinearity is referred to as *essential ill-conditioning*. A good example of this is in the correlation between gender and income in the current workforce. The fact that males earn more than their female counterparts creates this "ill-conditioning." However, when high correlation is due to an interaction variable, the multi-collinearity is referred to as *nonessential ill-conditioning*. [1]

155

When multi-collinearity exists, it is difficult to reliably assess the statistical significance, as well as the informal noticeable importance, of the highly correlated variables. Accordingly, multi-collinearity makes it difficult to define a strategy for excluding variables. A sizable literature has developed for essential ill-conditioning, which has produced several approaches for specifying models. [2] In contrast, there is a modest collection of articles for nonessential ill-conditioning. [3,4]

8.2 Strategy for Modeling with Interaction Variables

The popular strategy for modeling with interaction variables is the *Principle of Marginality*, which states that a model including an interaction variable should also include the component variables that define the interaction. [5,6] A cautionary note accompanies this principle: *neither testing the statistical significance (or noticeable importance) nor interpreting the coefficients of the component variables should be performed.* [7] A significance test, which requires a unique partitioning of the dependent variable variance in terms of the interaction variable and its components, is not possible due to the multi-collinearity.

An unfortunate by-product of this principle is that models with *unnecessary* component variables go undetected. Such models are prone to overfit, which results in either unreliable predictions and/or deterioration of performance, as compared to a well-fit model with the necessary component variables.

An alternative strategy is based on Nelder's *Notion of a Special Point*. [8] Nelder's strategy relies on understanding the functional relationship between the component variables and the dependent variable. When theory or prior knowledge about the relationship is limited or unavailable, Nelder suggests using exploration data analysis to determine the relationship. However, Nelder provides no general procedure or guidelines for uncovering the relationship among the variables.

I propose using CHAID as the exploratory method for the uncovering the functional relationship among the variables. Higher order interactions are seldom found to be significant, at least in database marketing. Therefore, I will limit this discussion to only first-order interactions. If higher order interactions are required, the proposed method can be extended.

8.3 Strategy Based on the Notion of a Special Point

For simplicity, consider the full model in Equation (8.1):

$$Y = b_0 + b_1X_1 + b_2X_2 + b_3X_1X_2 \tag{8.1}$$

$X_1 = 0$ is called a *special point* on the scale if when $X_1 = 0$ there is no relationship between Y and X_2. If $X_1 = 0$ is a special point, then omit X_2 from the full model; otherwise, X_2 should not be omitted.

Similarly for X_2. $X_2 = 0$ is called a *special point* on the scale if when $X_2 = 0$ there is no relationship between Y and X_1. If $X_2 = 0$ is a special point, then omit X_1; otherwise, X_1 should not be omitted.

If both X_1 and X_2 have special points, then X_1 and X_2 are omitted from the full model, and the model is reduced to $Y = b_0 + b_3X_1X_2$.

If a zero value is not assumed by the component variable, then no special point exists and the procedure is not applicable.

8.4 Example of a Response Model with an Interaction Variable

Database marketing provides an excellent example for this alternative method. I illustrate the alternative with a response model case study, but it applies as well to a profit model. A music continuity club requires a model to increase response to its solicitations. Based on a sample random (size 299,214) of a recent solicitation with a 1% response, I conduct a logistic regression analysis of RESPONSE on two available predictor variables, X_1 and X_2. The variables are defined as:

1. RESPONSE is the indicator of a response to the solicitation: 0 indicates nonresponse, 1 indicates response.

2. X_1 is the number of months since the last inquiry. A zero month value indicates an inquiry was made within the month of the solicitation.

3. X_2 is a measure of an individual's affinity to the club based on the number of previous purchases and the listening interest categories of the purchases.

The output of the logistic analysis is presented in Table 8.1. X_1 and X_2 have Wald chi-square values 10.5556 and 2.9985, respectively. Using Wald cut-off value 4, as outlined in Chapter 2, Section 2.11.1, there is indication that X_1 is an important predictor variable, whereas X_2 is not quite important. The classification accuracy of the base RESPONSE model is displayed in Classification Table 8.2. The *Total* column represents the *actual* number of non-responders and responders: there are 296,120 nonresponders and 3,094 responders. The *Total* row represents the predicted or *classified* number of

TABLE 8.1

Logistic Regression of Response on X_1 and X_2

Variable	df	Parameter Estimate	Standard Error	Wald Chi-Square	Pr > Chi-Square
INTERCEPT	1	−4.5414	0.0389	13613.7923	0.E+00
X_1	1	−0.0338	0.0104	10.5556	0.0012
X_2	1	0.0145	0.0084	2.9985	0.0833

TABLE 8.2

Classification Table of Model with X_1 and X_2

		Classified		
		0	1	Total
Actual	0	143,012	153,108	296,120
	1	1,390	1,704	3,094
Total		144,402	154,812	299,214
			TCCR	48.37%

nonresponders and responders: there are 144,402 individuals classified as nonresponders; and 154,812 classified as responders. The diagonal cells indicate the model's correct classifications. The upper-left cell (actual = 0 and classified = 0) indicates the model correctly classified 143,012 non-responders. The lower-right cell (actual = 1 and classified = 1) indicates the model correctly classified 1,704 responders. The total correct classification rate (TCCR) is equal to 48.4% (= [143012+1704]/299214).

After creating the interaction variable X_1X_2 (= X_1*X_2), I conduct another logistic regression analysis of RESPONSE with X_1, X_2 and X_1X_2; the output is in Table 8.3. I can look at the Wald chi-square values, but all I will see is that no direct statistical assessment is possible as per the cautionary note.

The classification accuracy of this full RESPONSE model is displayed in Classification Table 8.4. TCCR(X_1, X_2, X_1X_2) equals 55.64%, which represents a 15.0% improvement over the TCCR(X_1, X_2) of 48.37% for the model without the interaction variable. These two TCCR values provide a benchmark for

TABLE 8.3

Logistic Regression of Response on X_1, X_2 and X_1X_2

Variable	df	Parameter Estimate	Standard Error	Wald Chi-Square	Pr > Chi-Square
INTERCEPT	1	−4.5900	0.0502	8374.8095	0.E+00
X_1	1	−0.0092	0.0186	0.2468	0.6193
X_2	1	0.0292	0.0126	5.3715	0.0205
X_1X_2	1	−0.0074	0.0047	2.4945	0.1142

TABLE 8.4

Classification Table of Model with X_1, X_2 and X_1X_2

		Classified		Total
		0	1	
Actual	0	164,997	131,123	296,120
	1	1,616	1,478	3,094
Total		166,613	132,601	299,214
			TCCR	55.64%

assessing the effects of omitting — if possible under the notion of a special point — X_1 and/or X_2.

Can component variable X_1 or X_2 be omitted from the full RESPONSE model in Table 8.3? To omit a component variable, say X_2, it must be established that there is no relationship between RESPONSE and X_2, when X_1 is a special point. CHAID can be used to determine whether or not the relationship exists. Following a brief review of the CHAID technique, I will illustrate how to use CHAID for this new approach.

8.5 CHAID for Uncovering Relationships

CHAID is a technique that recursively partitions (or splits) a population into separate and distinct segments. These segments, called nodes, are split in such a way that the variation of the dependent variable (categorical or continuous) is minimized within the segments and maximized among the segments. After the initial splitting of the population into two or more nodes (defined by values of an independent or predictor variable), the splitting process is repeated on each of the nodes. Each node is treated like a new sub-population. It is then split into two or more nodes (defined by the values of another predictor variable) such that the variation of the dependent variable is minimized within the nodes, and maximized among the nodes. The splitting process is repeated until stopping rules are met. The output of CHAID is a *tree* display, where the root is the population, and the branches are the connecting segments such that the variation of the dependent variable is minimized within all the segments, and maximized among all the segments.

CHAID was originally developed as a method of finding interaction variables. In database marketing, CHAID is primarily used today as a market segmentation technique. Here, I utilize CHAID for uncovering the relationship among component variables and dependent variable to provide the information needed to test for a special point.

For this application of CHAID, the RESPONSE variable is the dependent variable, and the component variables are predictor variables X_1 and X_2. The CHAID analysis is forced to produce a tree, in which the initial splitting of the population is based on the component variable to be tested for a special point. One of the nodes must be defined by a zero value. Then, the "zero" node is split by the other component variable, producing response rates for testing for a special point.

8.6   Illustration of CHAID for Specifying a Model

Can X_1 be omitted from the full RESPONSE model? If when $X_2 = 0$ there is no relationship between RESPONSE and X_1, then $X_2 = 0$ is a special point, and X_1 can be omitted from the model. The relationship between RESPONSE and X_1 can be easily assessed by the RESPONSE CHAID tree, in Figure 8.1, which is read and interpreted as follows.

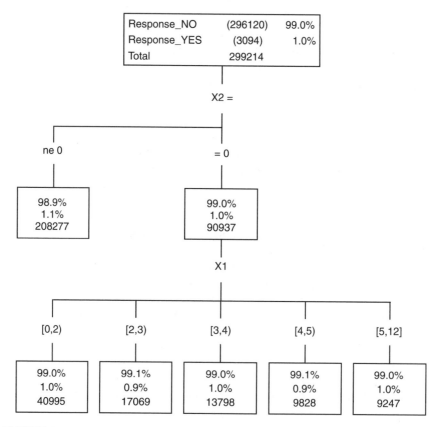

FIGURE 8.1
CHAID Tree for Testing X2 for a Special Point

1. The top box (root node) indicates that for the sample of 299,214 individuals, there are 3,094 responders, and 296,120 nonresponders. The response rate is 1% and nonresponse rate is 99%.

2. The left-leaf node (of the first level) of the tree represents 208,277 individuals whose X_2 values are not equal to zero. The response rate among these individuals is 1.1%.

3. The right-leaf node represents 90,937 individuals whose X_2 values equal zero. The response rate among these individuals is 1.0%.

4. Tree notation: Trees for a continuous predictor variable denote the continuous values in intervals: a closed interval, or a left-closed/right-open interval. The former is denoted by [a, b] indicating all values between and including a and b. The latter is denoted by [a, b) indicating all values greater than or equal to a, and less than b.

5. I reference the bottom row of five branches (defined by the intersection of X_2 and X_1 intervals/nodes), from left to right: #1 through #5.

6. The branch #1 represents 40,995 individuals whose X_2 values equal zero *and* X_1 values lie in [0, 2). The response rate among these individuals is 1.0%.

7. The branch #2 represents 17,069 individuals whose X_2 values equal zero *and* X_1 values lie in [2, 3). The response rate among these individual is 0.9%

8. The branch #3 represents 13,798 individuals whose X_2 values equal zero *and* X_1 values lie in [3, 4). The response rate among these individuals is 1.0%.

9. The branch #4 represents 9,828 individuals whose X_2 values equal zero *and* X_1 values lie in [4, 5). The response rate among these individual is 0.9%.

10. The node #5 represents 9,247 individuals whose X_2 values equal zero *and* X_1 values lie in [5, 12]. The response rate among these individuals is 1.0%.

11. The pattern of response rates (1.0%, 0.9%, 1.0%, 0.9%, 1.0%) across the five branches reveals there is no relationship between RESPONSE and X_1 when $X_2 = 0$. Thus, $X_2 = 0$ is a special point.

12. The implication is X_1 can be omitted from the RESPONSE model.

Can X_2 be omitted from the full RESPONSE model? If when $X_1 = 0$ there is no relationship between RESPONSE and X_2, then $X1 = 0$ is a special point, and X_2 can be omitted from the model. The relationship between RESPONSE and X_2 can be assessed by the CHAID tree in Figure 8.2, which is read and interpreted as follows:

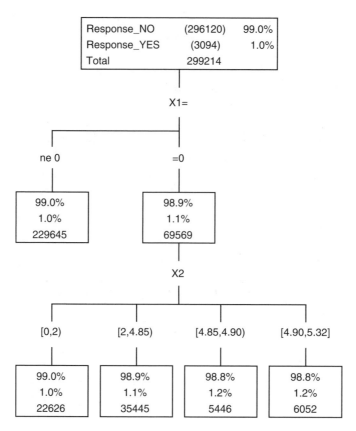

FIGURE 8.2
CHAID Tree for Testing X1 for a Special Point

1. The top box indicates that for the sample of 299,214 individuals there are 3,094 responders and 296,120 nonresponders. The response rate is 1% and nonresponse rate is 99%.
2. The left-leaf node represents 229,645 individuals whose X_1 values are not equal to zero. The response rate among these individuals is 1.0%.
3. The right-leaf node represents 69,569 individuals whose X_1 values equal zero. The response rate among these individuals is 1.1%.
4. I reference the bottom row of four branches (defined by the intersection of X_1 and X_2 intervals/nodes), from left to right: #1 through #4.
5. The branch #1 represents 22,626 individuals whose X_1 values equal zero *and* X_2 values lie in [0, 2). The response rate among these individuals is 1.0%.

6. The branch #2 represents 35,445 individuals whose X_1 values equal zero *and* X_2 values lie in [2, 4.85). The response rate among these individuals is 1.1%.

7. The branch #3 represents 5,446 individuals whose X_1 values equal zero *and* X_2 values lie in [4.85, 4.90). The response rate among these individuals is 1.2%.

8. The branch #4 represents 6,052 individuals whose X_1 values equal zero *and* X_2 values lie in [4.90, 5.32]. The response rate among these individuals is 1.2%.

9. The pattern of response rates across the four nodes is best observed in the smooth plot of response rates by the minimum values[1] of the intervals for the X_2-branches in Figure 8.3. There appears to be a positive straight-line relationship between RESPONSE and X_2 when $X_1 = 0$. Thus, $X_1 = 0$ is *not* a special point.

10. The implication is X_2 *cannot* be omitted from the RESPONSE model.

Because I choose to omit X_1, I performed a logistic regression analysis on RESPONSE with X_2 and X_1X_2; the output is in Table 8.5. The classification accuracy of this RESPONSE model is displayed in the Classification Table 8.6: TCCR(X_2, X_1X_2) equals 55.64%. TCCR for the full model, TCCR(X_1, X_2, X_1X_2), is also equal to 55.64%. The implication is that X_1 is not needed in the model.

In sum, the parsimonious (best-so-far) RESPONSE model is defined by X_2 and X_1X_2. As an alternative method for specifying a model with interaction variables, using CHAID justifies the omission of the component variables in situations like the one illustrated.

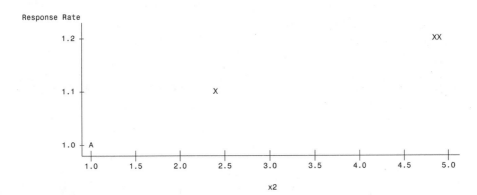

FIGURE 8.3
Smooth Plot of Response and X2

---

[1] The minimum value is one of several values which can be used; alternatives are the means or medians of the predefined ranges.

TABLE 8.5

Logistic Regression of Response on X_2 and X_1X_2

Variable	df	Parameter Estimate	Standard Error	Wald Chi-Square	Pr > Chi-Square
INTERCEPT	1	−4.6087	0.0335	18978.8487	0.E+00
X_1X_2	1	−0.0094	0.0026	12.6840	0.0004
X_2	1	0.0332	0.0099	11.3378	0.0008

TABLE 8.6

Classification Table of Model with X_2 and X_1X_2

		Classified		
		0	1	Total
Actual	0	165,007	131,113	296,120
	1	1,617	1,477	3,094
Total		166,624	132,590	299,214
			TCCR	55.64%

8.7 An Exploratory Look

A closer look at the plot in Figure 8.3 seems to indicate that relationship between RESPONSE and X_2 bends slightly in the upper-right portion, which would imply there is a quadratic component in the relationship. Because trees are always exploratory in nature, I choose to test the X_2 squared term ($X_2{*}X_2$), denoted by X_2_SQ, in the model. (Note: the Bulging Rule discussed in Chapter 3, which seeks to straighten unconditional data, does not apply here because the relationship between RESPONSE and X_2 is a conditional one, as it is based on individuals with the "condition $X_1 = 0$.")

The logistic regression analysis on RESPONSE with X_2, X_1X_2 and X_2_SQ is in Table 8.7. The classification accuracy of this model is shown in Classification Table 8.8. TCCR(X_2, X_1X_2, X_2_SQ) equals 64.59%, which represents a 16.1% improvement over the best-so-far model with TCCR(X_2, X_1X_2) equals 55.64%.

TABLE 8.7

Logistic Regression of Response on X_2, X_1X_2 and X_2_SQ

Variable	df	Parameter Estimate	Standard Error	Wald Chi-Square	Pr > Chi-Square
INTERCEPT	1	−4.6247	0.0338	18734.1156	0.E+00
X_1X_2	1	−0.0075	0.0027	8.1118	0.0040
X_2	1	0.7550	0.1151	43.0144	5.E-11
X_2_SQ	1	−0.1516	0.0241	39.5087	3.E-10

TABLE 8.8

Classification Table of Model with X_2, X_1X_2 and X_2_SQ

		Classified		
		0	1	Total
Actual	0	191,998	104,122	296,120
	1	1,838	1,256	3,094
Total		193,836	105,378	299,214
			TCCR	64.59%I

I conclude that the relationship is quadratic, and the corresponding model is a good fit of the data. Thus, the best RESPONSE model is defined by X_2, X_1X_2 and X_2_SQ.

8.8 Database Implication

Database marketers are among those who use response models to identify individuals most likely to respond to their solicitations, and thus place more value in the information in cell (1, 1) — the number of responders correctly classified — than in the TCCR. Table 8.9 indicates the number of responders correctly classified for the models tested. The model that *appears* to be the best is actually not the best for a database marketer because it identifies the least number of responders (1,256).

I summarize the modeling process: the base RESPONSE model with the two original variables, X_1 and X_2, produces TCCR(X_1, X_2) = 48.37%. The interaction variable X_1X_2, which is added to the base model, produces the full model with TCCR(X_1, X_2, X_1X_2) = 55.64% for a 14.9% classification improvement over the base model.

TABLE 8.9

Summary of Model Performance

Model			Number of Responder Correct	
Type	Defined by	TCCR	Classification	RCCR
base	X_1, X_2	48.37%	1,704	1.10%
full	X_1, X_2, X_1X_2	55.64%	1,478	1.11%
best-so-far	X_2, X_1X_2	55.64%	1,477	1.11%
best	X_2, X_1X_2, X_2_SQ	64.59%	1,256	1.19%

Using the new CHAID approach to determine whether a component variable can be omitted, I observe that X_2 (but not X_1) can be dropped from the full model. Thus, the best-so-far model — with X_2 and X_1X_2 — has no loss of performance over the full model: $TCCR(X_2, X_1X_2) = TCCR(X_1, X_2, X_1X_2) = 55.64\%$.

A closer look at the smooth plot in RESPONSE and X_2 suggests that X_2_SQ be added to the best-so-far model, producing the best model with a $TCCR(X_2, X_1X_2, X_2$_SQ$)$ of 64.59%, which indicates a 16.2% classification improvement over the best-so-far model.

Database marketers assess the performance of a response model by how well the model correctly classifies responders *among the total number of individuals classified as responders.* That is, the percent of responders correctly classified, or the *responder correct classification rate* (RCCR) is the pertinent measure. For the base model, the *Total* row in Table 8.2 indicates the model classifies 154,812 individuals as responders, among which there are 1,704 correctly classified: RCCR is 1.10% in Table 8.5. RCCR values for the best, best-so-far, and full models are 1.19%, 1.11% and 1.11%, respectively, in Table 8.5. Accordingly, the best RCCR-based model is still the best model, which was originally based on TCCR.

It is interesting to note that the performance improvement based on RCCR is not as large as the improvement based on TCCR. The best model compared to the best-so-far model has a 7.2% (= 1.19%/1.11%) RCCR-improvement vs. 16.2% (= 64.6%/55.6%) TCCR-improvement.

8.9 Summary

After briefly reviewing the concepts of interaction variables and multi-collinearity, and the relationship between the two, I restated the popular strategy for modeling with interaction variables: the Principle of Marginality states that a model including an interaction variable should also include the component variables that define the interaction. I reinforced the cautionary note that accompanies this principle: neither testing the statistical significance (or noticeable importance) nor interpreting the coefficients of the component variables should be performed. Moreover, I pointed out that an unfortunate by-product of this principle is that models with unnecessary component variables go undetected, resulting in either unreliable predictions and/or deterioration of performance.

Then I presented an alternative strategy, which is based on Nelder's Notion of a Special Point. I defined the strategy for first-order interaction, $X_1{}^*X_2$, as higher order interactions are seldom found in database marketing applications: predictor variable $X_1 = 0$ is called a special point on the scale if when $X_1 = 0$ there is no relationship between the dependent variable and a second predictor variable X_2. If $X_1 = 0$ is a special point, then omit X_2 from the model;

otherwise X_2 should not be omitted. I proposed using CHAID as the exploratory method for determining whether or not there is a relationship between the dependent variable and X_2.

I presented a case study involving the building of a database response model to illustrate the Special Point-CHAID alternative method. The resultant response model, which omitted one component variable, clearly demonstrated the utility of the new method. Then, I took advantage of the full-bodied case study, as well as the mantra of data mining (never stop digging into the data) to improve the model. I determined that the additional term, the square of the included component variable, added 16.2% improvement over the original response model in terms of the traditional measure of model performance, the total correct classification rate (TCCR).

Digging a little deeper, I emphasized the difference between the traditional and database measures of model performance. Database marketers are more concerned about models with a larger responder correct classification rate (RCCR) than a larger TCCR. As such, the improved response model, which initially appeared not to have the largest RCCR, was the best model in terms of RCCR as well as TCCR.

References

1. Marquardt, D.W., You should standardize the predictor variables in your regression model. *Journal of the American Statistical Association*, 75, 87–91, 1980.
2. Aiken, L.S. and West, S.G., *Multiple Regression: Testing and Interpreting Interactions*, Sage Publications, Thousand Oaks, CA ,1991.
3. Chipman, H., Bayesian variable selection with related predictors, *Canadian Journal of Statistics*, 24, 17–36, 1996.
4. Peixoto, J.L., Hierarchical variable selection in polynomial regression models, *The American Statistician*, 41, 311–313, 1987.
5. Nelder, J.A., Functional marginality is important (letter to editor), *Applied Statistics*, 46, 281–282, 1997.
6. McCullagh, P.M. and Nelder, J.A., *Generalized Linear Models*, Chapman & Hall, London, 1989.
7. Fox, J., *Applied Regression Analysis, Linear Models, and Related Methods*, Sage Publications, Thousand Oaks, CA, 1997.
8. Nelder, J., The selection of terms in response-surface models — how strong is the weak-heredity principle? *The American Statistician*, 52, 4, 1998.

9

Market Segment Classification Modeling with Logistic Regression[1]

Logistic regression analysis is a recognized technique for classifying individuals into two groups. Perhaps less known but equally important, polychotomous logistic regression (PLR) analysis is another method for performing classification. The purpose of this chapter is to present PLR analysis as a multi-group classification technique. I will illustrate the technique using a cellular phone market segmentation study to build a market segment classification model as part of a customer relationship management strategy, better known as CRM.

I start the discussion by defining the typical two-group (binary) logistic regression model. After introducing necessary notation for expanding the binary logistic regression model, I define the PLR model. For readers uncomfortable with such notation, the PLR model provides several equations for classifying individuals into one of many groups. The number of equations is one less than the number of groups. Each equation looks like the binary logistic regression model.

After a brief review of the estimation and modeling processes used in polychotomous logistic regression, I illustrate PLR analysis as a multi-group classification technique with a case study based on a survey of cellular phone users. The survey data was used initially to segment the cellular phone market into four groups. I use PLR analysis to build a model for classifying cellular users into one of the four groups.

9.1 Binary Logistic Regression

Let Y be a binary dependent variable that assumes two outcomes or classes, typically labeled 0 and 1. The binary logistic regression model (BLR) classifies an individual into one of the classes based on the values for predictor

---

[1] This chapter is based on an article with the same title in *Journal of Targeting Measurement and Analysis for Marketing*, 8, 1, 1999. Used with permission.

(independent) variables X_1, X_2, ..., X_n for that individual. BLR estimates the *logit of Y* — a log of the odds of an individual belonging to class 1; the logit is defined in Equation (9.1). The logit can easily be converted into the probability of an individual belonging to class 1, Prob(Y = 1), which is defined in Equation (9.2).

$$\text{logit } Y = b_0 + b_1 * X_1 + b_2 * X_2 + ... + b_n * X_n \tag{9.1}$$

$$\text{Prob}(Y = 1) = \frac{\exp(\text{logit } Y)}{1 + \exp(\text{logit } Y)} \tag{9.2}$$

An individual's predicted probability of belonging to class 1 is calculated by "plugging-in" the values of the predictor variables for that individual in Equations (9.1) and (9.2). The bs are the logistic regression coefficients, which are determined by the calculus-based method of maximum likelihood. Note that unlike the other coefficients, b_0 (referred to as the Intercept) has no predictor variable with which it is multiplied. Needless to say, the probability of an individual belonging to class 0 is 1 − Prob(Y = 1).[2]

9.1.1   Necessary Notation

I introduce notation that is needed for the next section. There are several explicit restatements (in Equations (9.3), (9.4), (9.5) and (9.6)) of the logit of Y of Equation (9.1). They are superfluous when Y takes on only two values, 0 and 1:

$$\text{logit } Y = b_0 + b_1 * X_1 + b_2 * X_2 + ... + b_n * X_n \tag{9.3}$$

$$\text{logit}(Y = 1) = b_0 + b_1 * X_1 + b_2 * X_2 + ... + b_n * X_n \tag{9.4}$$

$$\text{logit}(Y = 1 \text{ vs. } Y = 0) = b_0 + b_1 * X_1 + b_2 * X_2 + ... + b_n * X_n \tag{9.5}$$

$$\text{logit}(Y = 0 \text{ vs. } Y = 1) = - [b_0 + b_1 * X_1 + b_2 * X_2 + ... + b_n * X_n] \tag{9.6}$$

Equation (9.3) is the standard notation for the BLR model; it is assumed that Y takes on two values 1 and 0, and class 1 is the outcome being modeled. Equation (9.4) indicates that class 1 is being modeled, and assumes class 0 is the other category. Equation (9.5) formally states that Y has two classes, and class 1 is being modeled. Equation (9.6) indicates that class 0 is the outcome being modeled; this expression is the negative of the other expressions, as indicated by the negative sign on the right-hand side of the Equation.

---

[2] Because Prob (Y=0) + Prob (Y=1) = 1.

9.2   Polychotomous Logistic Regression Model

When the class-dependent variable takes on more than two outcomes or classes, the polychotomous logistic regression model, an extension of BLR model, can be used to predict class membership. For ease of presentation, I discuss Y with three categories, coded 0, 1 and 2.

Three binary logits can be constructed as in Equations (9.7), (9.8) and (9.9).[3]

$$\text{logit\_10} = \text{logit}(Y = 1 \text{ vs. } Y = 0) \tag{9.7}$$

$$\text{logit\_20} = \text{logit}(Y = 2 \text{ vs. } Y = 0) \tag{9.8}$$

$$\text{logit\_21} = \text{logit}(Y = 2 \text{ vs. } Y = 1) \tag{9.9}$$

I use the first two logits (because of the similarity with the standard expression of the BLR) to define the PLR model in Equations (9.10), (9.11) and (9.12):

$$\text{Prob}(Y = 0) = \frac{1}{1 + \exp(\text{logit\_10}) + \exp(\text{logit\_20})} \tag{9.10}$$

$$\text{Prob}(Y = 1) = \frac{\exp(\text{logit\_10})}{1 + \exp(\text{logit\_10}) + \exp(\text{logit\_20})} \tag{9.11}$$

$$\text{Prob}(Y = 2) = \frac{\exp(\text{logit\_20})}{1 + \exp(\text{logit\_10}) + \exp(\text{logit\_20})} \tag{9.12}$$

The PLR model is easily extended when there are more than three classes. When Y = 0, 1, ..., k (i.e., k+1 outcomes) the model is defined in Equations (9.13), (9.14) and (9.15):

$$\text{Prob}(Y = 0) = \frac{1}{1 + \exp(\text{logit\_10}) + \exp(\text{logit\_20}) + ... + \exp(\text{logit\_k0})} \tag{9.13}$$

$$\text{Prob}(Y = 1) = \frac{\exp(\text{logit\_10})}{1 + \exp(\text{logit\_10}) + \exp(\text{logit\_20}) + ... + \exp(\text{logit\_k0})} \tag{9.14}$$

...,

$$\text{Prob}(Y = k) = \frac{\exp(\text{logit\_k0})}{1 + \exp(\text{logit\_10}) + \exp(\text{logit\_20}) + ... + \exp(\text{logit\_k0})} \tag{9.15}$$

---

[3] It can be shown that from any pair of logits, the remaining logit can be obtained.

where
  logit_10 = logit(Y = 1 vs. Y = 0),
  logit_20 = logit(Y = 2 vs. Y = 0),
  logit_30 = logit(Y = 3 vs. Y = 0),
  ...,
  logit_k0 = logit(Y = k vs. Y = 0).

Note: There are k logits for a PLR with k+1 classes.

9.3   Model Building with PLR

PLR is estimated by the same method used to estimate BLR, namely, maximum likelihood estimation. The theory for stepwise variable selection, model assessment and validation has been worked out for PLR. Some theoretical problems still remain. For example, a variable can be declared significant for all but, say, one logit. Because there is no theory for estimating a PLR model with the constraint of setting a coefficient equal to zero for a given logit, the PLR model may produce unreliable classifications.

Choosing the best set of predictor variables is the toughest part of modeling, and is perhaps more difficult with PLR because there are k logit equations to consider. The traditional stepwise procedure is the popular variable selection process for PLR. (See Chapter 16 for a discussion of the traditional stepwise procedure.) Without arguing the pros and cons of the stepwise procedure, its use as the determinant of the final model is questionable.[4] The stepwise approach is best as a rough-cut method for boiling down many variables — about 50 or more — to a manageable set of about 10. I prefer the CHAID methodology as the variable selection process for the PLR, as it fits well into the data mining paradigm of digging into the data to find unexpected structure. In the next section, I illustrate CHAID as the variable selection procedure for building a market segmentation classification model for a cellular phone.

9.4   Market Segmentation Classification Model

In this section, I describe the cellular phone user study. Using the four user groups derived from a cluster analysis of the survey data, I proceed to build a four-group classification model with PLR. I use CHAID to identify the final

---

[4] Briefly, stepwise is misleading because all possible subsets are not considered; the final selection is too data dependent, and sensitive to influential observations. Also, it does not automatically check for model assumptions; and does not automatically test for interaction terms. Moreover, the stepwise does not guarantee to find the globally best subset of the variables.

set of candidate predictor variables, interaction terms, and variable structures (i.e., re-expressions of original variables defined with functions such as log or square root) for inclusion in the model. After a detailed discussion of the CHAID analysis, I define the market classification model, and assess the total classification accuracy of the resultant model.

9.4.1 Survey of Cellular Phone Users

A survey of 2,005 past and current users of cellular phones from a wireless carrier was conducted to gain an understanding of customer needs, as well as the variables that effect churn (cancellation of cellular service) and long-term value. The survey data was used to segment this market of consumers into homogenous groups so that group-specific marketing programs, namely CRM strategies, can then be developed to maximize the individual customer relationship.

A cluster analysis was performed that produced four segments. Segment names and sizes are in Table 9.1. The Hassle-free segment is concerned with the ability of the customer to design the contract and rate plan. The Service segment is focused on quality of the call, such as no dropped calls and the clarity of calls. The Price segment values discounts, such as offering 10% off the base monthly charges and 30 free minutes of use. Last, the Features segment represents the latest technology, such as long-lasting batteries and free phone upgrades. A model is needed to divide the wireless carrier's entire database into these four actionable segments, after which marketing programs can be used in addressing the specific needs of these predefined groups.

The survey was appended with information from the carrier's billing records. For all respondents, who are now classified into one of four segments, there are ten usage variables, such as number of mobile phones, minutes of use, peak and off-peak calls, airtime revenue, base charges, roaming charges, and free minutes of use (yes/no).

TABLE 9.1

Cluster Analysis Results

Name	Size
Hassle-free	13.2% (265)
Service	24.7% (495)
Price	38.3% (768)
Features	23.8% (477)
Total	100% (2,005)

9.4.2   CHAID Analysis

Briefly, CHAID is a technique that recursively partitions a population into separate and distinct subpopulations or nodes such that the variation of the dependent variable is minimized within the nodes, and maximized among the nodes. The dependent variable can be binary (dichotomous), polychotomous or continuous. The nodes are defined by independent variables, which pass through an algorithm for partitioning. The independent variables can be categorical or continuous.

To perform a CHAID analysis I must define the dependent variable and the set of independent variables. For this application of CHAID, the set of independent variables is the set of usage variables appended to the survey data. The dependent variable is the class variable Y identifying the four segments from the cluster analysis. Specifically, I define the dependent variable as follows:

Y = 0 if segment is Hassle-free

   = 1 if segment is Service

   = 2 if segment is Price

   = 3 if segment is Features

The CHAID analysis identifies[5] four important predictor variables:[6]

1. NUMBER OF MOBILE PHONES — the number of mobile phones a customer has
2. MONTHLY OFF-PEAK CALLS — the number of off-peak calls averaged over a three-month period
3. FREE MINUTES, yes/no: free first 30 minutes of use per month
4. MONTHLY AIRTIME REVENUE, total revenue excluding monthly charges averaged over a three-month period

The CHAID tree for NUMBER OF MOBILE PHONES, shown in Figure 9.1, is read as follows:

1. The top box indicates that for the sample of 2,005 customers the sizes (and incidences) of the segments are 264 (13.2%), 495 (24.75), 767 (38.3%) and 479 (23.9%), for Hassle-free, Service, Price and Features segments, respectively.
2. The left node represents 834 customers who have one mobile phone. Within this subsegment, the incidence rates of the four segments are 8.5%, 21.8%, 44.7% and 24.9%, for Hassle-free, Service, Price and Features segments, respectively.

---

[5] Based on the average chi-square value per degrees of freedom (number of nodes).
[6] I do not consider interaction variables identified by CHAID because the sample was too small.

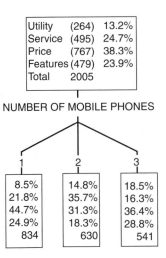

FIGURE 9.1
CHAID Tree for NUMBER OF MOBILE PHONES

3. The middle node represents 630 customers who have two mobile phones, with incidence rates of 14.8%, 35.7%, 31.3% and 18.3%, for Hassle-free, Service, Price and Features segments, respectively.

4. The right node represents 541 customers who have three mobile phones, with incidence rates of 18.5%, 16.3%, 36.4% and 28.8%, for Hassle-free, Service, Price and Features segments, respectively.

The CHAID tree for MONTHLY OFF-PEAK CALLS, shown in Figure 9.2, is read as follows:

1. The top box is the sample breakdown of the four segments; it is identical to the NUMBER OF MOBILE PHONES top box.

2. There are three nodes: the left node is defined by the number of calls in the half-open interval [0,1), which means zero calls; the middle node is defined by the number of calls in the half-open interval [1,2), which means one call; and the right node is defined by the number of calls in the closed interval [2, 270], which means calls between greater than/equal to 2 and less than/equal to 270.

3. The left node represents 841 customers who have zero off-peak calls. Within this subsegment, the incidence rates of the four segments are 18.1%, 26.4%, 32.6% and 22.9% for Hassle-free, Service, Price and Features, respectively.

4. The middle node represents 380 customers who have one off-peak call, with incidence rates of 15.3%, 27.1%, 43.9% and 13.7% for Hassle-free, Service, Price and Features, respectively.

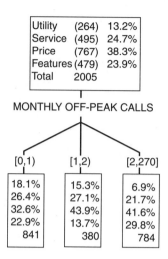

FIGURE 9.2
CHAID Tree for MONTHLY OFF-PEAK CALLS

5. The right node represents 784 customers who have off-peak calls inclusively between 2 and 270, with incidence rates of 6.9%, 21.7%, 41.6% and 29.8% for Hassle-free, Service, Price and Features, respectively.

Similar readings can be made for the other predictor variables, FREE MINUTES and MONTHLY AIRTIME REVENUE, identified by CHAID. Their CHAID trees are in Figures 9.3 and 9.4.

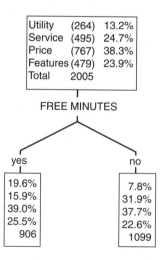

FIGURE 9.3
CHAID Tree for FREE MINUTES

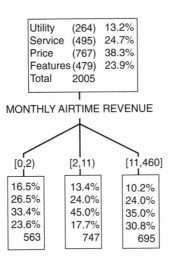

Utility	(264)	13.2%
Service	(495)	24.7%
Price	(767)	38.3%
Features	(479)	23.9%
Total	2005	

MONTHLY AIRTIME REVENUE

[0,2)	[2,11)	[11,460]
16.5%	13.4%	10.2%
26.5%	24.0%	24.0%
33.4%	45.0%	35.0%
23.6%	17.7%	30.8%
563	747	695

FIGURE 9.4
CHAID tree for MONTHLY AIRTIME REVENUE

Analytically, CHAID declaring a variable significant means the segment incidence rates (as a column array of rates) differ significantly across the nodes. For example, MONTHLY OFF-PEAK CALLS has three column arrays of segment incidence rates, corresponding to the three nodes, {18.1%, 26.4%, 32.6%, 22.9%}, {15.3%, 27.1%, 43.9%, 13.7%} and {6.9%, 21.7%, 41.6%, 29.8%}. These column arrays are significantly different from each other. This is a complex concept that really has no interpretive value, at least in the context of identifying variables with classification power. However, the CHAID tree can help evaluate the potential predictive power of variable when it is transformed into a tree graph.

9.4.3   CHAID Tree Graphs

Displaying a CHAID tree in a graph facilitates the evaluation of the potential predictive power of a variable. I plot the incidence rates by the minimum values[7] of the intervals for the nodes, and connect the *smooth* points to form a *trace line*, one for each segment. The shape of the trace line indicates the effect of the predictor variable on identifying individuals in a segment. The baseline plot, which indicates a predictor variable with no classification power, consists of all the segment trace lines being horizontal or "flat." The extent to which the segment trace lines are not flat indicates the potential predictive power of the variable for identifying an individual belonging to the segments. A comparison of all trace lines (one for each segment) provides a total view of how the variable effects classification *across* the segments.

---

[7] The minimum value is one of several values which can be used; alternatives are the mean or median of each predefined interval.

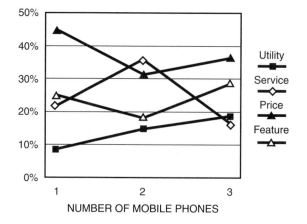

FIGURE 9.5
CHAID Tree Graph for NUMBER OF MOBILE PHONES

The following discussion relies on an understanding of the basics of re-expressing data as discussed in Chapter 3. Suffice to say that sometimes the predictive power offered by a variable can be increased by re-expressing or transforming the original form of the variable. The final forms of the four predictor variables identified by CHAID variables are given in the next section.

The PLR is a linear model,[8] which requires a linear or straight-line relationship between predictor variable and each implicit binary segment dependent variable.[9] The tree graph suggests the appropriate re-expression when the empirical relationships between predictor and binary segment variables are not linear.[10]

The CHAID tree graph for NUMBER OF MOBILE PHONES in Figure 9.5 indicates the following:

1. There is a positive and nearly linear relationship[11] between NUMBER OF MOBILE PHONES and the identification of customers in the Utility segment. This implies that only the variable NUMBER OF MOBILE PHONES in its raw form, no re-expression, may be needed.

2. The relationship for the Features segment also has a positive relationship but with bend from below.[12] This implies that the variable

---

[8] That is, each individual logit is a sum of weighted predictor variables.
[9] For example, for segment Hassle-free (Y=0); binary Hassle-free segment variable=1 if Y=0; otherwise Hassle-free segment variable=0.
[10] The suggestions are determined from the Ladder of Powers and the Bulging Rule discussed in Chapter 3.
[11] A relationship is assessed by determining the slope between the left and right node smooth points.
[12] The position of a bend is determined by the middle-node smooth point.

NUMBER OF MOBILE PHONES in its raw form, and its square may be required.

3. Price segment has a negative effect with a bend from below. This implies that the variable NUMBER OF MOBILE PHONES in itself in raw form, and its square root may be required.

4. Service segment has a negative relationship with a bend from above. This implies that the variable NUMBER OF MOBILE PHONES itself in raw form, and its square may be required.

The CHAID tree graphs for the other predictor variables, MONTHLY OFF-PEAK CALLS, MONTHLY AIRTIME REVENUE, AND FREE MINUTES are in Figures 9.6, 9.7 and 9.8. Interpreting the graphs, I diagnostically determine the following:

1. MONTHLY OFF-PEAK CALLS — this variable in its raw form, its square and square root forms may be required.

2. MONTHLY AIRTIME REVENUE — this variable in its raw form, and square and square root forms may be required.

3. FREE MINUTES — this variable as is, in its raw form may be needed.

The preceding CHAID tree graph analysis serves as the initial variable selection process for PLR. The final selection criterion for including a variable in the PLR model is that the variable must be significantly/noticeably important on no less than three of the four logit equations based on the techniques discussed in Chapter 3.

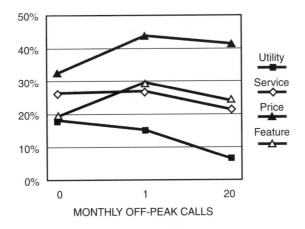

FIGURE 9.6
CHAID Tree Graph for MONTHLY OFF-PEAK CALLS

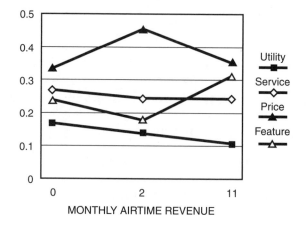

FIGURE 9.7
CHAID Tree Graph for MONTHLY AIRTIME REVENUE

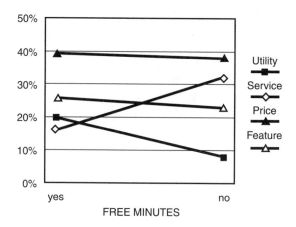

FIGURE 9.8
CHAID Tree Graph for FREE MINUTES

9.4.4 Market Segment Classification Model

The final PLR model for classifying customers into one of the four cellular phone behavioral market segments has the following variables:

1. Number of Mobile Phones (NMP)
2. Square of NMP
3. Square root of NMP
4. Monthly Off-Peak Calls
5. Monthly Airtime Revenue (MAR)
6. Square of MAR

7. Square root of MAR
8. Free Minutes

Without arguing the pros and cons of validation procedures, I draw a fresh holdout sample of size 5,000 to assess the total classification accuracy of the model.[13] The classification results of the model in Table 9.2 are read as follows:

1. The row totals are the *actual* counts in the sample. The sample consists of 650 Hassle-free customers, 1,224 Service customers, 1,916 Price customers and 1,210 Features customers. The percent figures are the percent compositions of the segments with respect to the total sample. For example, 13.0% of the sample consists of actual Hassle-free customers.

2. The column totals are *predicted* counts. The model predicts: 655 Hassle-free customers, 1,268 Service customers, 1,870 Price customers and 1,207 Features customers. The percent figures are the percent compositions with respect to the predicted counts. For example, the model predicts 13.1% of the sample as Hassle-free customers.

3. Given that the sample consists of 13.0% Hassle-free customers and the model predicts 13.1% Hassle-free customers, the model has no *bias* with respect to classifying Hassle-free. Similarly, the model shows no bias in classifying the other groups: for Service the actual incidence is 24.5% vs. the predicted 25.4%; for Price the actual incidence is 38.3% vs. the predicted 37.4%; and for Features the actual incidence is 24.2% vs. the predicted 24.1%.

TABLE 9.2

Market Segment Model: Classification Table of Cell-Counts

		PREDICTED				
		Hassle-free	Service	Price	Features	Total
	Hassle-free	326 (50.0%)	68	158	98	650 (13.0%)
ACTUAL	Service	79	460 (36.3%)	410	275	1,224 (24.5%)
	Price	147	431	922 (49.0%)	416	1,916 (38.3%)
	Features	103	309	380	418 (34.6%)	1,210 (24.2%)
	Total	655 (13.1%)	1,268 (25.4%)	1,870 (37.4%)	1,207 (24.1%)	5,000 (100%)

---

[13] Here is a great opportunity to debate and test various ways of calibrating and validating a 'difficult' model under the best of conditions.

4. Although the model shows no bias, the big question is, How accurate are the predictions? Among those customers predicted to be Hassle-free, how many actually are? Among those customers predicted to be Service, how many are actually Service? Similarly, for Price and Features. The percents in the table cells provide the answer. For Hassle-free, the model correctly classifies 50.0% (= 326/655) of the time. Without a model, I would expect 13.0% correct classifications of Hassle-free. Thus, the model has a lift of 385 (50.0%/13.0%), i.e., the model provides 3.85 times the number of correct classifications of Hassle-free customers obtained by chance.

5. The model has a lift of 148 (36.3%/24.5%) for Service; a lift of 128 (49.0%/38.3%) for Price; and a lift of 143 (34.6%/24.2%) for Features.

6. As a summary measure of how well the model makes correct classifications, I look at the total correct classification rate (TCCR). Simply put, TCCR is total number of correct classifications across all groups divided by total sample size. Accordingly, I have 326 + 460 + 922 + 418 = 2,126 divided by 5,000, which yields TCCR = 42.52%.

7. To assess the improvement of total correct classification provided by the model, I must compare it to the total correct classification provided by the model chance. TCCR(chance model) is defined as the sum of the squared actual group incidence. For the data at hand, TCCR(chance model) is 28.22% (= (13.0%*13.0%) + (24.5%*24.5%) + (38.3%*38.3%) + (24.2%*24.2%)).

8. Thus, the model lift is 151 (= 42.52%/28.22%). That is, the model provides 51% more total correct classifications across all groups than obtained by chance.

9.5   Summary

I cast the multi-group classification technique of polychotomous logistic regression (PLR) as an extension of the more familiar two-group (binary) logistic regression model. I derived the PLR model for a dependent variable assuming k+1 groups by "piecing together" k individual binary logistic models. I proposed CHAID as the variable selection process for the PLR, as it fits well into the data-mining paradigm to dig into the data to find important predictor variables and unexpected structure.

To bring into practice the PRL, I illustrated the building of a classification model based on a four-group market segmentation of cellular phone users. For the variable selection procedure, I demonstrated how CHAID can be used. For this application of CHAID, the dependent variable is the variable identifying the four market segments. CHAID trees were used to identify the starter subset of predictor variables. Then, the CHAID trees were transformed into CHAID tree graphs, which offer potential re-expressions — or identification of structures — of the starter variables. The final market segment classification model had a few re-expressed variables (involving square roots and squares).

Last, I assessed the performance of the final four-group/market segment classification model in terms of total correct classification rate (TCCR). TCCR for the final model was 42.52%, which represented a 51% improvement over the TCCR (28.22%) when no model is used to classify customers into one of the four market segments.

10

CHAID as a Method for Filling in Missing Values

The problem of analyzing data with missing values is well known to data analysts. Data analysts know that almost all standard statistical analyses require complete data for reliable results. These analyses performed with incomplete data assuredly produce biased results. Thus, data analysts make every effort to fill in the missing data values in their datasets. The popular solutions to the problem of handling missing data belong to the collection of imputation, or fill-in techniques. This chapter presents CHAID as an alternative method for filling in missing data.

10.1 Introduction to the Problem of Missing Data

Missing data are a pervasive problem in data analysis. It is the rare exception when the data at hand have no missing data values. The objective of filling in missing data is to recover or minimize the loss of information due to the incomplete data. Below, I provide a brief introduction to handling missing data.

Consider a random sample of ten individuals in Table 10.1 described by three variables, AGE, GENDER and INCOME. There are missing values, which are denoted by a dot (.). Eight out of ten individuals provide their age; seven out of ten individuals provide their gender and income.

Two common solutions to handling missing data are *available-case analysis* and *complete-case analysis*.[1] The available-case analysis uses *only the cases* for which the variable of interest is available. Consider the calculation of the mean AGE: the available sample size (number of nonmissing values) is eight, not the original sample size of ten. The calculation for the means of INCOME and GENDER[2] uses two different available samples of size seven. The cal-

---

[1] Available-case analysis is also known as pairwise deletion. Complete-case analysis is also known as listwise deletion or casewise deletion.

[2] The mean of GENDER is the incidence of females in the sample.

TABLE 10.1

Random Sample of Ten Individuals

Individual	Age (years)	Gender (0 = male; 1 = female)	Income
1	35	0	$50,000
2	.	.	$55,000
3	32	0	$75,000
4	25	1	$100,000
5	41	.	.
6	37	1	$135,000
7	45	.	.
8	.	1	$125,000
9	50	1	.
10	52	0	$65,000
Total	317	4	$605,000
Number of Nonmissing Values	8	7	7
Mean	39.6	57%	$86,429

'.' = missing value

culation on different samples points to a weakness in available-case analysis. Unequal sample sizes create practical problems. Comparative analysis across variables is difficult because different sub-samples of the original sample are used. Also, estimates of multi-variable statistics are prone to illogical values.[3]

The popular complete-case analysis uses examples for which *all variables* are present. A complete-case analysis of the original sample in Table 10.1 includes only five cases as reported in Table 10.2. The advantage of this type

TABLE 10.2

Complete-Case Version

Individual	Age (years)	Gender (0 = male; 1 = female)	Income
1	35	0	$50,000
3	32	0	$75,000
4	25	1	$100,000
6	37	1	$135,000
10	52	0	$65,000
Total	181	2	$425,000
Number of Nonmissing Values	5	5	5
Mean	36.2	40%	$85,000

---

[3] Consider the correlation coefficient of X_1 and X_2. If the available sample sizes for X_1 and X_2 are unequal, it is possible to obtain a correlation coefficient value that lies outside the theoretical [-1,1] range.

of analysis is simplicity, because standard statistical analysis can be applied without modification for incomplete data. Comparative analysis across variables is not complicated because only one common sub-sample of the original sample is used. The disadvantage of discarding incomplete cases is the resulting loss of information.

Another solution is *dummy variable adjustment*. [1] For a variable X with missing data, two new variables are used in its place. X_filled and X_dum are defined as follows:

1. X_filled = X, if X is not missing; X_filled = 0 if X is missing.
2. X_dum = 0 if X is not missing; X_dum = 1 if X is missing.

The advantage of this solution is its simplicity of use without having to discard cases. The disadvantage is that the analysis can become unwieldy when there are a lot of variables with missing data. In addition, filling in the missing value with a zero is arbitary, which is unsettling for some data analysts.

Among the missing-data solutions is the imputation method, which is defined as any process that fills in missing data to produce a complete data set. The simplest and most popular imputation method is *mean-value imputation*. The mean of the nonmissing values of the variable of interest is used to fill in the missing data. Consider individuals 2 and 8 in Table 10.1. Their missing ages are replaced with the mean AGE of the file, namely, 40 years (rounded from 39.6). The advantage of this method is undoubtedly its ease of use. The means may be calculated within classes, which are predefined by other variables related to the study at hand.

Another popular method is *regression-based imputation*. Missing values are replaced by the predicted values from a regression analysis. The dependent variable Y is the variable whose missing values need to be imputed. The predictor variables, the Xs, are the *matching variables*. Y is regressed on the Xs using a complete-case analysis dataset. If Y is continuous, then ordinary least-squares (OLS) regression is appropriate. If Y is categorical, then the logistic regression model (LRM) is used. For example, I wish to impute AGE for individual 8 in Table 10.1. I Regress AGE on GENDER and INCOME (the matching variables) based on the complete-case dataset consisting of five individuals (ID #s: 1, 3, 4, 6 and 10). The OLS regression imputation model is defined in Equation (10.1):

$$AGE\_imputed = 25.8 - 20.5*GENDER + 0.0002*INCOME \qquad (10.1)$$

Plugging in the values of GENDER (= 1) and INCOME (= $125,000) for individual 8, the imputed AGE is 53 years.

10.2 Missing-Data Assumption

Missing-data methods presuppose that the missing data are "missing at random." Rubin formalized this condition into two separate assumptions. [2]

1. Missing at random (MAR) means that what is missing does not depend on the missing values, but may depend on the observed values.

2. Missing completely at random (MCAR) means that what is missing does not depend on either the observed values or the missing values. When this assumption is satisfied for all variables, the reduced sample of individuals with only complete data can be regarded as a simple random sub-sample from the original data. Note that the second assumption MCAR represents a stronger condition than the MAR assumption.

The missing-data assumptions are problematic. The MCAR assumption can be tested to some extent by comparing the information from complete cases to the information from incomplete cases. A procedure often used is to compare the distribution of the variable of interest, say, Y, based on non-missing data with the distribution of Y based on missing data. If there are significant differences, then the assumption is considered not met. If no significant differences are indicated, then the test offers no direct evidence of assumption violation. In this case, the assumption can be considered to have been cautiously met.[4] The MAR assumption is impossible to test for validity. (Why?)

It is accepted wisdom that missing-data solutions at best perform satisfactorily, even when the amount of missing data is moderate and the missing-data assumption has been met. The potential of the new two imputation methods, *maximum likelihood* and *multiple imputation*, which offers substantial improvement over the complete-case analysis, is questionable as their assumptions are easily violated. Moreover, their utility in big data applications has not been established.

Nothing can take the place of the missing data. Allison notes that "the best solution to the missing data problem is not to have any missing data." [3] Dempster and Rubin warn that "imputation is both seductive and dangerous," seductive because it gives a false sense of confidence that the data are complete and dangerous because it can produce misguided analyses and untrue models. [4]

The above admonitions are without reference to the impact of big data on filling in missing data. In big data applications, the problem of missing data is severe, as it is not uncommon for at least one variable to have 30%-90%

---

[4] This test proves the necessary condition for MCAR. It remains to be shown that there is no relationship between missingness on a given variable and the values of that variable.

of their values missing. Thus, I strongly argue that the imputation of big data applications must be used with restraint, and their findings must be used judiciously.

In the spirit of the EDA tenet — that failure is when you fail to try — I advance the proposed CHAID imputation method as an hybrid mean-value/regression-based imputation method that explicitly accommodates missing data without imposing additional assumptions. The salient features of the new method are the EDA characteristics:

1. Flexibility — assumption-free CHAID work especially well with big data containing large amounts of missing data
2. Practicality — descriptive CHAID tree provides analysis of data
3. Innovation — CHAID algorithm defines imputation classes
4. Universality — blending of two diverse traditions: traditional imputation methods, and machine learning algorithm for data-structure identification
5. Simplicity — CHAID tree imputation estimates are easy to use

10.3 CHAID Imputation

I introduce a couple of terms required for the discussion of CHAID imputation. Imputation methods require the sample to be divided into groups or classes, called *imputation classes*, which are defined by variables called matching variables. The formation of imputation classes is an important step to insure the reliability of the imputation estimates. As the homogeneity of the classes increases, so does the accuracy and stability of the estimates. It is assumed that the variance (with respect to the variable whose missing values to be imputed) within each class is small.

CHAID is a technique that recursively partitions a population into separate and distinct groups, which are defined by the predictor variables, such that the variance of the dependent variable is minimized within the groups, and maximized across the groups. CHAID was originally developed as a method of detecting 'combination' or interaction variables. In database marketing today, CHAID primarily serves as a market segmentation technique. Here, I propose CHAID as an alternative method for mean-value/regression-based imputation.

The justification for CHAID as a method of imputation is as follows: by definition, CHAID creates optimal homogenous groups, which can be used effectively as *trusty* imputation classes.[5] Accordingly, CHAID provides a reliable method of mean-value/regression-based imputation.

The CHAID methodology provides the following:

---

[5] Some analysts may argue about the optimality of the homogeneous groups but not the trustworthiness of the imputation classes.

CHAID is a tree-structured, assumption-free modeling alternative to OLS regression. It provides reliable estimates without the assumption of specifying the true structural form of the model, i.e., knowing the correct independent variables and their correct re-expressed forms, and without regard for the weighty classical assumptions of underlying OLS model. Thus, CHAID with its trusty imputation classes provides reliable *regression tree* imputation for a continuous variable.

CHAID can be used as a nonparametric tree-based alternative to the binary and polychotomous LRM without the assumption of specifying the true structural form of the model. Thus, CHAID with its trusty imputation classes, provides reliable *classification tree* imputation for a categorical variable.

CHAID potentially offers more reliable imputation estimates due to its ability to use most of the analysis sample. The analysis sample for CHAID is not as severely reduced by the pattern of missing values in the matching variables, as is the case for regression-based models, because CHAID can accommodate missing values for the matching variables in its analysis.[6] The regression-based imputation methods cannot make such an accommodation.

10.4 Illustration

Consider a sample of 29,132 customers from a cataloguer's database. The following is known about the customers: their ages (AGE_CUST), GENDER, total lifetime dollars (LIFE_DOL) and whether a purchase was made within the past three months (PRIOR_3). Missing values for each variable are denoted by ' ???. '

The counts and percentages of missing and nonmissing values for the variables are in Table 10.3. For example, there are 691 missing values for

TABLE 10.3

Counts and Percentages of Missing and Nonmissing Values

Variable	Missing		Nonmissing	
	Count	%	Count	%
AGE_CUST	1,142	3.9	27,990	96.1
GENDER	2,744	9.4	26,388	90.6
LIFE_DOL	691	2.4	28,441	97.6
PRIOR_3	965	3.3	28,167	96.7
All Variables	2,096	7.2	27,025	92.8

---

[6] Missing values are allowed to "float" within the range of the matching variable and rest at the position that optimizes the homogeneity of the groups.

LIF_DOL resulting in 2.4% missing rate. It is interesting to note that the complete-case sample size for an analysis with all four variables is 27,025. This represents a 7.2% (= 2,107/29,132) loss of information from discarding incomplete cases from the original sample.

10.4.1 CHAID Mean-Value Imputation for a Continuous Variable

I wish to impute the missing values for LIFE_DOL. I perform a mean-value imputation with CHAID using LIFE_DOL as the dependent variable and AGE_CUST as the predictor (matching) variable. The AGE_CUST CHAID tree is in Figure 10.1.

I set some conventions to simplify the discussions of the CHAID analyses.

1. The left-closed/right-opened interval [x, y) indicates values between x and y, including x and excluding y.
2. The closed interval [x, y] indicates values between x and y, including both x and y.
3. The distinction is made between nodes and imputation classes. Nodes are the visual displays of the CHAID groups. Imputation classes are defined by the nodes.
4. Nodes are referenced by numbers (1, 2, ...) from left to right as they appear in the CHAID tree.

The AGE_CUST CHAID tree, in Figure 10.1, is read as follows.

1. The top box indicates a mean LIFE_DOL of $27,288.47 for the available sample of 28,441 (nonmissing) observations for LIFE_DOL.
2. The CHAID creates four nodes with respect to AGE_CUST. Node 1 consists of 6,499 individuals whose ages are in the interval [18, 30) with a mean LIFE_DOL of $14,876.75. Node 2 consists of 7,160 individuals whose ages are in the interval [30, 40) with a mean

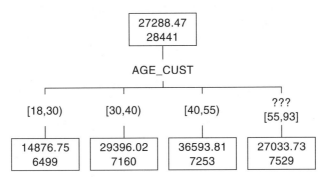

FIGURE 10.1
AGE_CUST CHAID Tree for LIF_DOL

LIFE_DOL of $29,396.02. Node 3 consists of 7,253 individuals whose ages are in the interval [40, 55) with a mean LIFE_DOL of $36,593.81. Node 4 consists of 7,529 individuals whose ages are either in the interval [55, 93] or missing. The mean LIFE_DOL is $27,033.73.

3. Note: CHAID positions the missing values of AGE_CUST in the "oldest-age missing" node.

The set of AGE_CUST CHAID mean-value imputation estimates for LIFE_DOL is the mean-values of the nodes 1 to 4: $14,876.75, $29,396.02, $36,593.81 and $27,033.73, respectively. The AGE_CUST distribution of the 691 individuals with missing LIFE_DOL is in Table 10.4. All the 691 individuals belong to the last class, and their imputed LIFE_DOL value is $27,033.73. Of course, if any of the 691 individuals were in the other AGE_CUST classes, the corresponding mean-values would be used.

10.4.2 Many Mean-Value CHAID Imputations for a Continuous Variable

CHAID provides many mean-value imputations — as many as there are matching variables — as well as a measure to determine which imputation estimates to use. The "goodness" of a CHAID tree is assessed by the measure percent variance explained (PVE). The imputation estimates based on the matching variable with the largest PVE value are often selected as the preferred estimates. Note, however, a large PVE value does not necessarily guarantee reliable imputation estimates, and the largest PVE value does not necessarily guarantee the best imputation estimates. The data analyst may have to perform the analysis at hand with imputations based on several of the large PVE-value matching variables.

Continuing with imputation for LIFE_DOL, I perform two additional CHAID mean-value imputations. The first CHAID imputation, in Figure 10.2, uses LIFE_DOL as the dependent variable and GENDER as the matching variable. The second imputation, in Figure 10.3, uses LIFE_DOL as the dependent variable and PRIOR_3 as the matching variable. The PVE values for the matching variables AGE_CUST, GENDER and PRIOR_3 are 10.20%, 1.45% and 1.52%, respectively. Thus, the preferred imputation estimates for

TABLE 10.4

CHAID Imputation Estimates For Missing Values of LIFE_DOL

AGE_CUST Class	Class Size	Imputation Estimate
[18, 30)	0	$14,876.75
[30,40)	0	$29,396.02
[40, 55)	0	$36,593.81
[55, 93] or ???	691	$27,033.73

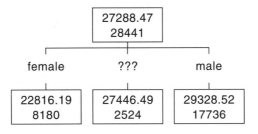

FIGURE 10.2
GENDER CHAID Tree for LIF_DOL

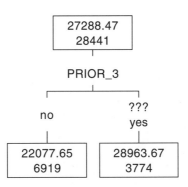

FIGURE 10.3
PRIOR_3 CHAID Tree for LIF_DOL

LIF_DOL are based on AGE_CUST, because the AGE_CUST-PVE value is noticeably the largest.

Comparative Note: Unlike CHAID mean-value imputation, traditional mean-value imputation provides no guideline for selecting an all-case continuous matching variable (e.g., AGE_CUST), whose imputation estimates are preferred.

10.4.3 Regression-Tree Imputation for LIF_DOL

I can selectively add matching variables[7] to the preferred single-variable CHAID tree — generating a regression tree — to increase the reliability of the imputation estimates, i.e., increase the PVE value. Adding GENDER and PRIOR_3 to the AGE_CUST tree, I obtain a PVE value of 12.32%, which represents an increase of 20.8% (= 2.12%/10.20%) over the AGE_CUST-PVE value. The AGE_CUST-GENDER-PRIOR_3 regression tree is displayed in Figure 10.4

---

[7] Here is a great place to discuss the relationship between increasing the number of variables in a (tree) model and its effects on bias and stability of the model's estimates.

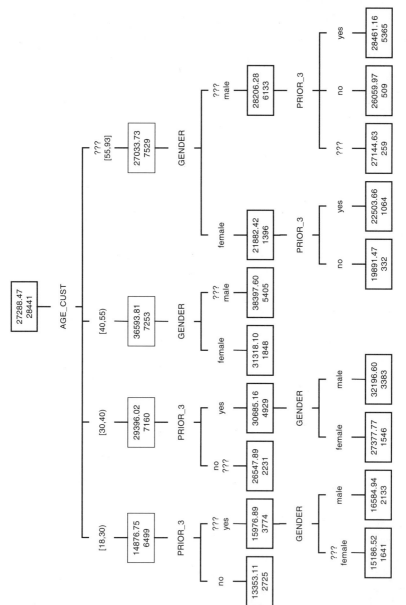

FIGURE 10.4

AGE_CUST, GENDER and PRIOR_3 Regression Tree for LIFE_DOL

The AGE_CUST-GENDER-PRIOR_3 regression tree is read as follows.

1. Extending the AGE_CUST CHAID tree, I obtain a regression tree with thirteen end nodes.
2. Node 1 (two levels deep) consists of 2,725 individuals whose ages are in the interval [18, 30) *and* have not made a purchase in the past three months (PRIOR_3 = no). The mean LIFE_DOL is $13,352.11.
3. Node 2 (three levels deep) consists of 1,641 individuals whose ages are in the interval [18, 30) *and* PRIOR_3 = ??? or yes *and* whose GENDER = ??? or female. The mean LIFE_DOL is $15,186.52.
4. The remaining nodes are interpreted similarly.

The AGE_CUST-GENDER-PRIOR_3 regression-tree imputation estimates for LIFE_DOL are the mean-values of thirteen end nodes. An individual's missing LIFE_DOL value is replaced with the mean-value of the imputation class to which the individual matches. The distribution of the 691 individuals with missing LIFE_DOL values in terms of the three matching variables is in Table 10.5. All the missing LIFE_DOL values come from the five right-most end nodes (Nodes 9 to 14). As before, if any of the 691 individuals were in the other eight nodes, the corresponding mean-values would be used.

Comparative Note: Traditional OLS regression-based imputation for LIF_DOL based on four matching variables, AGE_CUST, two dummy variables for GENDER ("missing" GENDER is considered a category), and one dummy variable for PRIOR_3, results in a complete-case sample size of 27,245, which represents a 4.2% (= 1,196/28,441) loss of information from the CHAID analysis sample.

TABLE 10.5

Regression Tree Imputation Estimates for Missing Values of LIFE_DOL

AGE_CUST Class	GENDER	PRIOR_3	Class Size	Imputation Estimates
??? or [55, 93]	Female	No	55	$19,891.47
??? or [55, 93]	Female	Yes	105	$22,503.66
??? or [55, 93]	Male	No	57	$26,059.97
??? or [55, 93]	Male	Yes	254	$28,461.16
??? or [55, 93]	???	No	58	$26,059.97
??? or [55, 93]	???	Yes	162	$28,461.16
Total			691	

10.5 CHAID Most-Likely Category Imputation for a Categorical Variable

CHAID for imputation of a categorical variable is very similar to the CHAID with a continuous variable, except for slight changes in assessment and interpretation. CHAID with a continuous variable assigns a mean-value to an imputation class. In contrast, CHAID with a categorical variable assigns the predominate or most-likely category to an imputation class. CHAID with a continuous variable provides PVE. In contrast, with a categorical variable, CHAID provides the measure Proportion of Total Correct Classifications (PTCC)[8] to identify the matching variable(s) whose imputation estimates are preferred.

As noted in CHAID with a continuous variable, there is a similar note that a large PTCC value does not necessarily guarantee reliable imputation estimates, and the largest PTCC value does not necessarily guarantee the best imputation estimates. The data analyst may have to perform the analysis with imputations based on several of the large PTCC-value matching variables.

10.5.1 CHAID Most-Likely Category Imputation for GENDER

I wish to impute the missing values for GENDER. I perform a CHAID most-likely category imputation using GENDER as the dependent variable and AGE_CUST as the matching variable. The AGE_CUST CHAID tree is in Figure 10.5. The PTCC value is 68.7%.

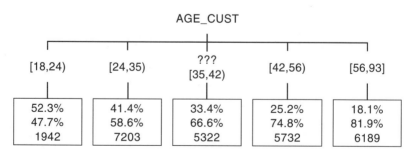

FIGURE 10.5
AGE_CUST CHAID Tree for GENDER

---

[8] PTCC is calculated with the percentage of observations in each of the end nodes of the tree that fall in the modal category. The weighted sum of these percentages over all end nodes of the tree is PTCC. A given node is weighted by the number of observations in the node relative to the total size of the tree.

The AGE_CUST CHAID tree is read as follows:

1. The top box indicates that the incidences of females and males are 31.6% and 68.4%, respectively, based on the available sample of 26,388 (nonmissing) observations for GENDER.
2. The CHAID creates five nodes with respect to AGE_CUST. Node 1 consists of 1,942 individuals whose ages are in interval [18, 24). The incidences of female and male are 52.3% and 47.7%, respectively. Node 2 consists of 7,203 individuals whose ages are in interval [24, 35). The incidences of female and male are 41.4% and 58.6%, respectively.
3. The remaining nodes are interpreted similarly.
4. Note: CHAID places the missing values in the "middle-age" node. Compare the LIFE_DOL CHAID: the missing ages are placed in the "oldest-age/missing" node.

I perform two additional CHAID most-likely category imputations, in Figures 10.6 and 10.7, for GENDER using the individual matching variables PRIOR_3 and LIF_DOL , respectively. The PTCC values are identical, 68.4%. Thus, I select the imputation estimates for GENDER based on AGE_CUST, as its PTCC value is the largest (68.7%), albeit not noticeably different from the other PTCC values.

The AGE_CUST CHAID most-likely category imputation estimates for GENDER are the most likely, that is, the largest percentage categories of the nodes in Figure 10.5: female (52.3%), male (58.6%), male (66.6%), male (74.8%), male (82.9%). I replace an individual's missing GENDER value with the predominate category of the imputation class to which the individual matches. There are 2,744 individuals with missing GENDER values whose AGE_CUST distribution is in Table 10.6. The individuals whose ages are in

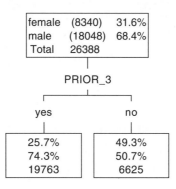

FIGURE 10.6
PRIOR_3 CHAID Tree for GENDER

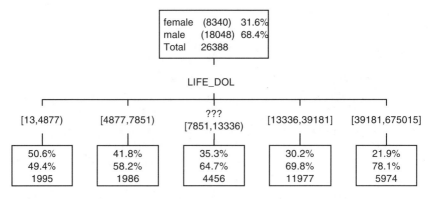

FIGURE 10.7
LIFE_DOL CHAID Tree for GENDER 10.8 AGE_CUST-PRIOR_3-LIF_DOL Classification Tree for GENDER

TABLE 10.6

CHAID Imputation Estimates for Missing Values of GENDER

AGE_CUST Class	Class Size	Imputation Estimate
[18, 24)	182	Female
[24, 35)	709	Male
??? or [35, 42)	628	Male
[42, 56)	627	Male
[56, 93]	598	Male
Total	2,744	

the interval [18, 24) are classified as female because the females have the largest percentage; all other individuals are classified as male. It is not surprising that most of the classifications are males, given the large (68.4%) incidence of males in the sample.

Comparative Note: Traditional mean-imputation can be performed with only matching variable PRIOR_3. (Why?) CHAID most-likely category imputation conveniently offers three choices of imputation estimates, and a guideline to select the best.

10.5.2 Classification Tree Imputation for GENDER

I can selectively add matching variables[9] to the preferred single-variable CHAID tree — generating a classification tree — to increase the reliability of the imputation estimates, i.e., increase the PTCC value. Extending the AGE_CUST tree, I obtain the classification tree, in Figure 10.8, for GENDER

---

[9] Here again is a great place to discuss the relationship between increasing the number of variables in a (tree) model and its effects on bias and stability of the model's estimates.

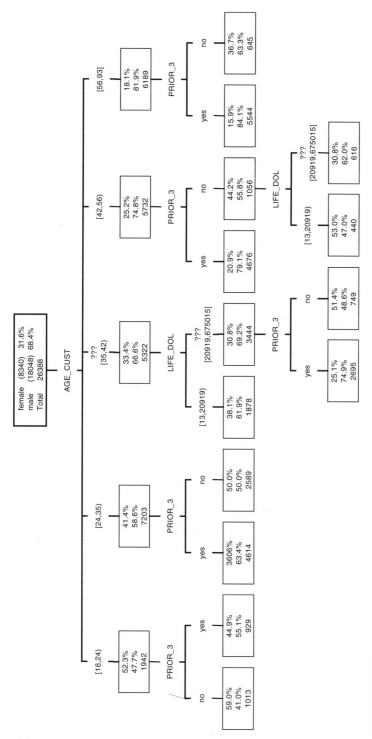

FIGURE 10.8

AGE_CUST-PRIOR_3-LIF_DOL Classification Tree for GENDER

based on AGE_CUST, PRIOR_3 and LIFE_DOL. The PTCC value for this tree is 79.3%, which represents an increase of 15.4% (= 10.6%/68.7%) over the AGE_CUST-PTCC value.

The AGE_CUST-PRIOR_3-LIFE_DOL classification tree is read as follows.

1. Extending the GENDER tree with the addition of PRIOR_3 and LIFE_DOL, I obtain a classification tree with twelve end nodes.

2. Node 1 consists of 1,013 individuals whose ages in the interval [18, 24) *and* who hav *not* made a prior purchase in the past three months (PRIOR_3 = no). The female and male incidences are 59.0% and 41.0%, respectively.

3. Node 2 consists of 929 individuals whose ages are in the interval [18, 24) *and* who have made a prior purchase in the past three months (PRIOR_3 = yes). The female and male incidences are 44.9% and 55.1%, respectively.

4. The remaining nodes are interpreted similarly.

5. Node 4 has no predominate category, as the GENDER incidences are equal to 50%.

6. Nodes 1, 7 and 9 have female as the predominate category; all remaining nodes have male as the predominate category.

The AGE_CUST-PRIOR_3-LIFE_DOL classification-tree imputation estimates for GENDER are the most-likely categories of the nodes. I replace an individual's missing GENDER value with the predominate category of the imputation class to which the individual belongs. The distribution of missing GENDER values, in Table 10.7, falls within all twelve nodes. Individuals in nodes 1, 7, and 9 are classified as females. For individuals in node 4: I flip a coin. All other individuals are classified as males.

Comparative Note: Traditional logistic regression-based imputation for GENDER based on three matching variables, AGE_CUST, LIF_DOL, and one dummy variable for PROIR_3, results in a complete-case sample size of 26,219, which represents a barely noticeable 0.6% (= 169/26,388) loss of information from the CHAID analysis sample.

10.6 Summary

It is rare to find data that have no missing data values. A given is that the data analyst first tries to recover or minimize the loss of information from the incomplete data. I briefly illustrated the popular missing-data methods, which include complete-case and available-case analyses, and mean-value and regression-based imputation methods. All these methods have at least one version of the missing-data assumptions: missing at random, and

TABLE 10.7

Classification Tree Imputation Estimates for Missing Values of GENDER

Node	AGE_CUST Class	PRIOR_3	LIF_DOL Class	Class Size	Imputation Estimates
1	[18, 24)	No	–	103	Females
2	[18, 24)	Yes	–	79	Males
3	[24, 35)	Yes	–	403	Males
4	[24, 35)	No	–	306	Females/ Males
5	??? or [35, 42)	–	[13, 20919)	169	Males
6	??? or [35, 42)	Yes	??? or [20919, 675015]	163	Males
7	??? or [35, 42)	No	??? or [20919, 675015]	296	Females
8	[42, 56)	Yes	–	415	Males
9	[42, 56)	No	[13, 20919)	70	Females
10	[42, 56)	No	[20919, 675015]	142	Males
11	[56, 93]	Yes	–	449	Males
12	[56, 93]	No	–	149	Males
Total				2,744	

missing completely at random, which are difficult and impossible to test for validity, respectively.

I remarked that the conventional wisdom of missing-data solutions is that their performance is, at best, satisfactory, especially for big data applications. Experts in missing data admonish us to remember that imputation is seductive and dangerous; therefore, the best solution of the missing data problem is not to have missing data. I strongly urged that imputation of big data applications must be used with restraint, and their findings must be used judiciously.

Then I recovered from the naysayers to advance the proposed CHAID imputation method. I presented CHAID as an alternative method for mean-value/regression-based imputation. The justification of CHAID for missing data is that CHAID creates optimal homogenous groups that can be used as trusty imputation classes, which insure the reliability of the imputation estimates. This renders CHAID as a reliable method of mean-value/regression-based imputation. Moreover, the CHAID imputation method has salient features commensurate with the best of what EDA offers.

I illustrated the CHAID imputation method with a database catalogue case study. I showed how CHAID — for both a continuous and categorical variable to be imputed — offers imputations based on several individual matching variables, and a rule for selecting the preferred CHAID mean-value imputation estimates. Also, I showed how CHAID — for both a continuous and categorical variable to be imputed — offers imputations based on a set of matching variables, and a rule for selecting the preferred CHAID regression-tree imputation estimates.

References

1. Cohen, J. and Cohen, P., *Applied Multiple Regression and Correlation Analysis for the Behavioral Sciences*, Erlbaum, Hillsdale, NJ, 1987.
2. Rubin, D.B., Inference and missing data, *Biometrika*, 63, 581–592, 1976.
3. Allison, P.D., *Missing Data*, Sage Publication, Thousand Oaks, CA, 2002, 2.
4. Dempster, A.P. and Rubin, D.B., Overview, in *Incomplete Data in Sample Surveys, Vol. II: Theory and Annotated Bibliography*, W.G. Madow, I. Okin, and D.B. Rubin, Eds., Academic Press, New York, 3–10, 1983.

11

Identifying Your Best Customers: Descriptive,
Predictive and Look-Alike Profiling[1]

Database marketers typically attempt to improve the effectiveness of their campaigns by targeting their best customers. Unfortunately, many database marketers are unaware that typical target methods develop a descriptive profile of their target customer — an approach that often results in less-than-successful campaigns. The purpose of this chapter is to illustrate the inadequacy of the *descriptive* approach and to demonstrate the benefits of the correct *predictive profiling* approach. I explain the predictive profiling approach, and then expand the approach to Look-Alike profiling.

11.1 Some Definitions

It is helpful to have a general definition of each of the three concepts discussed in this chapter. Descriptive profiles report the characteristics of a group of individuals. These profiles *do not allow* for drawing inferences about the group. The value of a descriptive profile lies in its definition of the target group's salient characteristics, which are used to develop an effective marketing strategy.

Predictive profiles report the characteristics of a group of individuals. These profiles *do allow* for drawing inferences about a specific behavior such as response. The value of a predictive profile lies in its predictions of the behavior of individuals in a target group; the predictions are used in producing a list of likely responders to a database marketing campaign.

A Look-Alike profile is a predictive profile based on a group of individuals who look like the individuals in a target group. When resources do not allow for the gathering of information on a target group, a predictive profile built on a surrogate or "look-alike" group provides a viable approach for predicting the behavior of the individuals in the target group.

---

[1] This chapter is based on an article with the same title in *Journal of Targeting, Measurement and Analysis for Marketing*, 10, 1, 2001. Used with permission.

11.2 Illustration of a Flawed Targeting Effort

Consider a hypothetical test mailing to a sample of 1,000 individuals conducted by Cell-Talk, a cellular phone carrier promoting a new bundle of
phone features. Three hundred individuals responded, yielding a 30%
response rate. (The offer also included the purchase of a cellular phone for
individuals who do not have one, but now want one because of the attractive
offer.) Cell-Talk analyzed the responders and profiled them in Tables 11.1
and 11.2 by using variables GENDER and OWN_CELL (current cellular
phone ownership), respectively. Ninety percent of the 300 responders are
males, and fifty-five percent already own a cellular phone. Cell-Talk concluded the typical responder is a male and owns a cellular phone.

Cell-Talk plans to target the next "features" campaign to males and to
owners of cellular phones. The effort is sure to fail. The reason for the poor
prediction is that the profile of their best customers (responders) is descriptive, not predictive. That is, the descriptive responder profile describes
responders *without regard* to responsiveness, and therefore the profile does
not imply that the best customers as defined are responsive.[2]

*Using a descriptive profile for predictive targeting draws a false implication of
the descriptive profile.* In our example, the descriptive profile of "90% of the
responders are males" *does not* imply that 90% of males are responders, or

TABLE 11.1

Responders and Nonresponders Profiles Response Rates by GENDER

GENDER	Responders		Nonresponders		Response Rate %
	Count	%	Count	%	
Female	30	10	70	10	30
Male	270	90	630	90	30
Total	300	100	700	100	

TABLE 11.2

Responders and Nonresponders Profiles Response Rates by OWN_CELL

OWN_CELL	Responders		Nonresponders		Response Rate %
	Count	%	Count	%	
Yes	165	55	385	55	30
No	135	45	315	45	30
Total	300	100	700	100	

---

[2] A descriptive responder profile may also describe a typical nonresponder. In fact, this is the situation in Tables 11.1 and 11.2.

even that males are more likely to respond.[3] Additionally, "55% of the responders who own cellular phones" *does not* imply that 55% of cellular phone owners are responders, or even that cellular phone owners are more likely to respond.

The value of a descriptive profile lies in its definition of the best customers' salient characteristics, which are used to develop an effective marketing strategy. In the illustration, knowing that the target customer is a male and owns a cellular phone, I position the campaign offer with a man wearing a cellular phone on his belt, instead of a woman reaching for a cellular phone in her purse. Accordingly, a descriptive profile tells how to talk to the target audience. And, as will be seen in the next section, a predictive profile helps find the target audience.

11.3 Well-Defined Targeting Effort

A predictive profile describes responders *with regard* to responsiveness, that is, in terms of variables that discriminate between responders and non-responders. Effectively, the discriminating or predictive variables produce *varied* response rates, and imply an expectation of responsiveness. To clarify this, consider the response rates for GENDER in Table 11.1. The response rates for both males and females are 30%. Accordingly, GENDER does not discriminate between responders and nonresponders (in terms of responsiveness). Similar results for OWN_CELL are in Table 11.2.

Hence, GENDER and OWN_CELL have no value as predictive profiles. Cell-Talk's targeting of males and current cellular phone owners is expected to generate the average, or sample response rate of 30%. In other words, this profile in a targeting effort will not produce more responders than will a random sample.

I now introduce a new variable, CHILDREN, which hopefully has predictive value. CHILDREN is defined as "yes" if individual belongs to a household with children, and "no" if individual does not belong to a household with children. Instead of discussing CHILDREN using a tabular display (such as in Tables 11.1 and 11.2), I prefer the user-friendly visual display of *CHAID trees*.

Response rates are best illustrated by use of CHAID tree displays. I review the GENDER and OWN_CELL variables in the tree displays in Figures 11.1 and 11.2, respectively. From this point on, I will refer only to the tree in this discussion, underscoring the utility of a tree as a profiler and reducing the details of tree building to nontechnical summaries.

The GENDER tree in Figure 11.1 is read as follows:

---

[3] More likely to respond than a random selection of individuals.

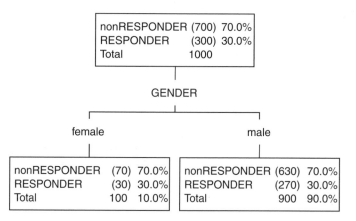

FIGURE 11.1
GENDER Tree

1. The top box indicates that for the sample of 1,000 individuals, there are 300 responders and 700 nonresponders. The response rate is 30% and nonresponse rate is 70%.
2. The left box represents 100 *females* consisting of 30 responders and 70 nonresponders. The response rate among the 100 females is 30%.
3. The right box represents 900 *males* consisting of 270 responders and 630 nonresponders. The response rate among the 900 males is 30%.

The OWN_CELL tree in Figure 11.2 is read as follows:

1. The top box indicates that for the sample of 1,000 individuals there are 300 responders and 700 nonresponders. The response rate is 30% and nonresponse rate is 70%.

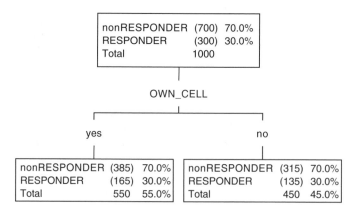

FIGURE 11.2
OWN_CELL Tree

2. The left box represents 550 individuals who *own* a cell phone. The response rate among these individuals is 30%.

3. The right box represents 450 individuals who *do not own* a cell phone. The response rate among these individuals is 30%.

The new variable CHILDREN is defined as presence of children in the household (yes/no). The CHILDREN tree in Figure 11.3 is read as follows:

1. The top box indicates that for the sample of 1,000 individuals there are 300 responders and 700 nonresponders. The response rate is 30% and nonresponse rate is 70%.

2. The left box represents 545 individuals belonging to households *with children*. The response rate among these individuals is 45.9%.

3. The right box represents 455 individuals belonging to households *with no children*. The response rate among these individuals is 11.0%.

CHILDREN has value as a predictive profile, as it produces varied response rates: 45.9% and 11.0%, for CHILDREN equal to "yes" and "no," respectively. If Cell-Talk targets individuals belonging to households with children, the expected response rate is 45.9%. This represents a *profile lift* of 153. (Profile lift is defined as profile response rate 45.9% divided by sample response rate 30%, multiplied by 100.) Thus, a targeting effort to the predictive profile is expected to produce 1.53 times more responses than expected from a random solicitation.

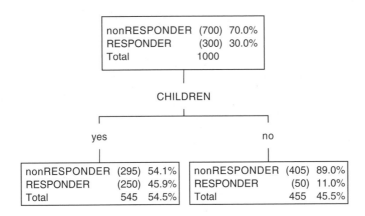

FIGURE 11.3
CHILDREN Tree

11.4 Predictive Profiles

Using additional variables, I can grow a single-variable tree into a full tree with many interesting and complex predictive profiles. Although actual building of a full tree is beyond the scope of this chapter, suffice to say, *a tree is grown to create end-node profiles (segments) with the greatest variation in response rates across all segments.* A tree has value as a *set of predictive profiles* to the extent 1) the number of segments with response rates greater than the sample response rate is "large" and 2) the corresponding profile (segment) lifts are "large." [4]

Consider the full tree defined by GENDER, OWN_CELL and CHILDREN in Figure 11.4. The tree is read as follows:

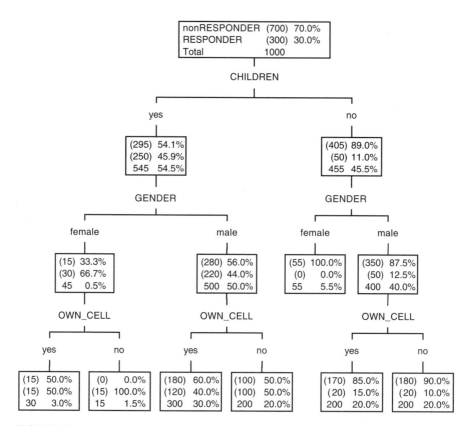

FIGURE 11.4
Full Tree defined by GENDER, OWN_CELL and CHILDREN

[4] The term "large" is subjective. Accordingly, the tree building process is subjective, which is an inherent weakness in CHAID trees.

1. The top box indicates that for the sample of 1,000 individuals, there are 300 responders and 700 nonresponders. The response rate is 30% and nonresponse rate is 70%.

2. I reference the end-node segments from left to right: #1 through #7.

3. The segment #1 represents 30 *females* who *own* a cellular phone and belong to households *with children*. The response rate among these individuals is 50.0%.

4. The segment #2 represents 15 *females* who *do not own* a cellular phone and belong to households *with children*. The response rate among these individuals is 100.0%.

5. The segment #3 represents 300 *males* who *own* a cellular phone and belong to households *with children*. The response rate among these individuals is 40.0%.

6. The segment #4 represents 200 *males* who *do not own* a cellular phone and belong to households *with children*. The response rate among these individuals is 50.0%.

7. The segment #5 represents 55 *females* who belong to households *with no children*. The response rate among these individuals is 0.0%.

8. The segment #6 represents 200 *males* who *own* a cellular phone and belong to households *with no children*. The response rate among these individuals is 15.0%.

9. The segment #7 represents 200 *males* who *do not own* a cellular phone and belong to households *with no children*. The response rate among these individuals is 10.0%.

I provide a summary of the segment response rates in the *gains chart* in Table 11.3. The construction and interpretation of the gains chart are as follows:

1. Segments are descendingly ranked by segment response rate.

2. In addition to the descriptive statistics (size of segment, number of responses, segment response rate), various calculated statistics are posted in the chart. They include the self-explanatory cumulative responses, segment response rate, and cumulative response rate.

3. The last statistic posted is the cumulative lift. Cumulative lift is defined as cumulative response rate divided by the sample response rate, multiplied by 100. It measures how many more responses are obtained by targeting various levels of aggregated segments over a random solicitation. Cumulative lift is discussed in detail below.

4. CHAID trees unvaryingly identify *sweet-spots* — segments with above-average response rates[5] that account for small percentages of

---

[5] Extreme small segment response rates (close to both 0% and 100%) reflect another inherent weakness of CHAID trees.

TABLE 11.3

Gains Chart for Tree Defined by GENDER, OWN CELL and CHILDREN

Segment*	Size of Segment	Number of Responses	Cumulative Responses	Segment Response Rate	Cumulative Rate	Cumulative Lift
#2 - OWN_CELL, no GENDER, female CHILDREN, yes	15	15	15	100.0%	100.0%	333
#1 - OWN_CELL, yes GENDER, female CHILDREN, yes	30	15	30	50.0%	66.7%	222
#3 - OWN_CELL, no GENDER, male CHILDREN, yes	200	100	130	50.0%	53.1%	177
#4 - OWN_CELL, yes GENDER, male CHILDREN, yes	300	120	250	40.0%	45.9%	153
#6 - OWN_CELL, yes GENDER, male CHILDREN, no	200	30	280	15.0%	37.6%	125
#7 - OWN_CELL, no GENDER, male CHILDREN, no	200	20	300	10.0%	31.7%	106
#5 - GENDER, female CHILDREN, no	55	0	300	0.0%	30.0%	100
	1,000	300		30.0%		

*Segments are ranked by response rates.

the sample. Under the working assumption that the sample is random and accurately reflects the population under study, sweet-spots account for small percentages of the population. There are two sweet-spots: segment #2 has a response rate of 100% and accounts for only 1.5% (= 15/1000) of the sample/population; segment #1 has a response rate of 50%, and accounts for only 3.0% (= 30/1000) of the sample/population.

5. A targeting strategy to a single sweet-spot is limited, as it is effective for only solicitations of mass products to large populations. Consider sweet-spot #2 with a population of 1,500,000. Targeting this segment produces a campaign of size 22,500 with an expected yield of 22,500 responses. Mass product-large campaigns have low break-even points, which make sweet-spot targeting profitable.

6. Reconsider sweet-spot #2 in a moderate size population of 100,000. Targeting this segment produces a campaign of size 1,500 with an expected yield of 1,500 responses. However, small campaigns for mass products have high break-even points, which render such campaigns neither practical nor profitable. In contrast, upscale product-small campaigns have low break-even points, which make sweet-spot targeting (e.g., to potential Rolls Royce owners) both practical and profitable.

7. For moderate size populations, the targeting strategy is to solicit an aggregate of several top consecutive responding segments to yield a campaign of a cost-efficient size to insure a profit. Here, I recommend a solicitation consisting of the top three segments, which would account for a 24.5% (= (15+30+200)/1000) of the population with an expected 53.1%. (See cumulative response rate for segment #3 row in Table 11.3.) Consider a population of 100,000. Targeting the aggregate of segments #2, #1 and #3 yields a campaign of size 24,500 with an expected 11,246 (= 53.1%*24,500) responses. Assuming a product offering with a good profit margin, the aggregate-targeting approach should be successful.

8. With respect to the cumulative lift, Cell-Talk can expect the following:

   a) Cumulative lift of 333 by targeting the top segment, which accounts for only 1.5% of the population. This is sweet-spot targeting as previously discussed.

   b) Cumulative lift of 222 by targeting the top two segments, which account for only 4.5% (= (15+30)/1000) of the population. This is effectively an enhanced sweet-spot targeting because the percentage of the aggregated segments is small.

   c) Cumulative lift of 177 by targeting the top three segments, which account for 24.5% of the population. This is the recommended targeting strategy as previously discussed.

d) Cumulative lift of 153 by targeting the top four segments, which account for 54.5% ((15+30+200+300)/1000) of the population. Unless the population is not too large, a campaign targeted to the top four segments may be cost prohibitive.

11.5 Continuous Trees

So far, the profiling uses only categorical variables, that is, variables that assume two or more discrete values. Fortunately, trees can accommodate continuous variables, or variables that assume many numerical values, which allows for developing profiles with both categorical and continuous variables.

Consider INCOME, a new variable. The INCOME tree in Figure 11.5 is read as follows:

1. Tree notation: Trees for a continuous variable denote the continuous values in ranges: a closed interval, or a left-closed/right-open interval. The former is denoted by [x, y] indicating all values between and including x and y. The latter is denoted by [x, y) indicating all values greater than or equal to x, and less than y.

2. The top box indicates that for the sample of 1,000 individuals there are 300 responders and 700 nonresponders. The response rate is 30% and nonresponse rate is 70%.

3. The segment #1 represents 300 individuals with income in the interval [$13,000, $75,000). The response rate among these individuals is 43.3%.

4. The segment #2 represents 200 individuals with income in the interval [$75,000, $157,000). The response rate among these individuals is 10.0%.

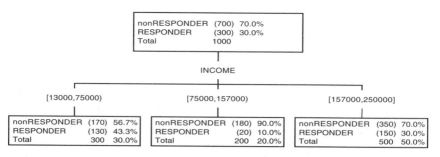

FIGURE 11.5
INCOME Tree

5. The segment #3 represents 500 individuals with income in the interval [$157,000, $250,000]. The response rate among these individuals is 30.0%.

The determination of the *number* of nodes and the *range* of an interval are based on a computer-intensive iterative process, which tests all possible numbers and ranges. That is, *a tree is grown to create nodes/segments with the greatest variation in response rates across all segments.*

The full tree with the variables GENDER, OWN_CELL, CHILDREN and INCOME is displayed in Figure 11.6. The gains chart of this tree is in Table 11.4. Cell-Talk can expect a cumulative lift of 177, which accounts for 24.5% of the population, by targeting the top three segments.

It is interesting to compare this tree, which includes INCOME, to the tree without INCOME in Figure 11.4. Based on Tables 11.3 and 11.4, these two trees have the same performance statistics, at least for the top three segments, a cumulative lift of 177, accounting for 24.5% of the population.

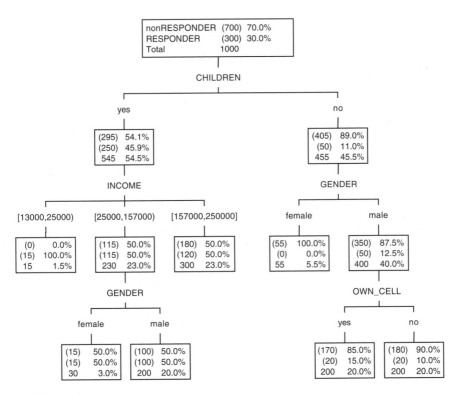

FIGURE 11.6
Full Tree Defined by GENDER, OWN_CELL, CHILDREN and INCOME

TABLE 11.4

Gains Chart for Tree Defined by GENDER, OWN CELL, CHILDREN and INCOME

Segment*	Size of Segment	Number of Responses	Cumulative Responses	Segment Response Rate	Cumulative Response Rate	Cumulative Lift
#1 - INCOME, [13000,25000) CHILDREN, yes	15	15	15	100.0%	100.0%	333
#3 - GENDER, male INCOME, [25000,157000) CHILDREN, yes	200	100	115	50.0%	53.5%	178
#2 - GENDER, female INCOME, [25000,157000) CHILDREN, yes	30	15	130	50.0%	53.1%	177
#4 - INCOME [157000,250000] CHILDREN, yes	300	120	250	40.0%	45.9%	153
#6 - OWN_CELL, yes GENDER, male CHILDREN, no	200	30	280	15.0%	37.6%	125
#7 - OWN_CELL, no GENDER, male CHILDREN, no	200	20	300	10.0%	31.7%	106
#5 - GENDER, female CHILDREN, no	55	0	300	0.0%	30.0%	100
	1,000	300		30.0%		

*Segments are ranked by response rates.

This raises some interesting questions. Does INCOME add any noticeable predictive power? In other words, how important is INCOME? Which tree is better? Which set of variables is best? There is a simple answer to these questions (and many more tree-related questions): *an analyst can grow many equivalent trees to explain the same response behavior. The tree that suits the analyst is the best (at least for that analyst).* Again, detailed answers to these questions (and more) are beyond the scope of this chapter.

11.6 Look-Alike Profiling

In the Cell-Talk illustration, Cell-Talk requires predictive profiles to increase response to its campaign for a new bundle of features. They conduct a test mailing to obtain a group of their best customers — responders of the new bundle offer — on which to develop the profiles.

Now, consider that Cell-Talk wants predictive profiles to target a solicitation based on a rental list of names, for which only demographic information is available, and the ownership of a cellular phone is not known. The offer is a discounted rate plan, which should be attractive to cellular phone owners with high monthly usage (around 500 minutes of use per month).

Even though Cell-Talk does not have the time or money to conduct another test mailing to obtain their *target group* of high-monthly-usage responders, they can still develop profiles to help in their targeting efforts, as long as they have a notion of what their target group looks like. Cell-Talk can use a *look-alike* group — individuals who look like individuals in the target group — as a substitute for the target group. This substitution allows Cell-Talk to develop Look-Alike profiles: profiles that identify individuals (in this case, persons on the rental list) who are most likely to look like individuals in the target group.

The construction of the look-alike group is important. The greater the similarity between the look-alike group and target group, the greater the reliability of the resultant profiles.

Accordingly, the definition of the look-alike group should be as precise as possible to insure the look-alikes are good substitutes for the target individuals. The definition of the look-alike group can include as many variables needed to describe pertinent characteristics of the target group. Note, the definition always involves at least one variable that is *not* available on the solicitation file, or in this case, the rental list. If all the variables are available, then there is no need for look-alike profiles!

Cell-Talk believes the target group looks like their current upscale cellular phone subscribers. Because cellular conversation is not inexpensive, Cell-Talk assumes that heavy-users must have high income to afford the cost of cellular use. Accordingly, Cell-Talk defines the look-alike group as individuals with a cellular phone (OWN_CELL = yes) *and* INCOME greater than $175,000.

Look-Alike profiles are based on the following assumption: individuals who look like individuals in a target group have levels of responsiveness similar to the group. Thus, the look-alike individuals serve as surrogates, or would-be responders.

Keep in mind that *individuals identified by Look-Alike profiles are expected probabilistically to look like the target group, but not expected necessarily to respond. In practice, it has been shown that the Look-Alike assumption is tenable, as solicitations based on Look-Alike profiles produce noticeable response rates.*

Look-Alike profiling via tree analysis identifies variables that discriminate between look-alike individuals, and nonlook-alike individuals (the balance of the population without the look-alike individuals). Effectively, the discriminating variables produce *varied* look-alike rates. I use the original sample data in Table 11.1 along with INCOME to create the LOOK-ALIKE variable required for the tree analysis. LOOK-ALIKE equals 1 if an individual has OWN_CELL = yes *and* INCOME greater than $175,000; otherwise, LOOK-ALIKE equals 0. There are 300 look-alikes and 700 nonlook-alikes, resulting in a sample look-alike rate of 30.0%.[6] These figures are reflected in the top box of the Look-Alike Tree in Figure 11.7. (Note that the sample look-alike rate and the original sample response rate are equal; this is purely coincidental.)

The gain chart for the Look-Alike tree is in Table 11.5. Targeting the top segment yields a cumulative lift of 333 (with a 30% depth of population). This means the predictive look-alike profile is expected to identify 3.33 times more individuals who look like the target group than expected from random selection (of 30% of rented names).

A closer look at the Look-Alike tree raises a question. INCOME is used in defining both the LOOK-ALIKE variable and the profiles. Does this indicate that the tree is poorly defined? No. For this particular example, INCOME is a wanted variable. Without INCOME, this tree could not guarantee that the identified *males with children* have high incomes, a requirement for being a look-alike.

11.7 Look-Alike Tree Characteristics

It is instructive to discuss a noticeable characteristic of Look-Alike trees. In general, upper segment rates in a Look-Alike tree are quite large and often reach 100%. Similarly, lower segment rates are quite small, and often fall to 0%. I observe these patterns in Table 11.5. There is one segment with a 100% look-alike rate, and three with a 0% look-alike rates.

---

[6] The sample look-alike rate sometimes needs to be adjusted to equal the incidence of the target group in the population. This incidence is rarely known and must be estimated.

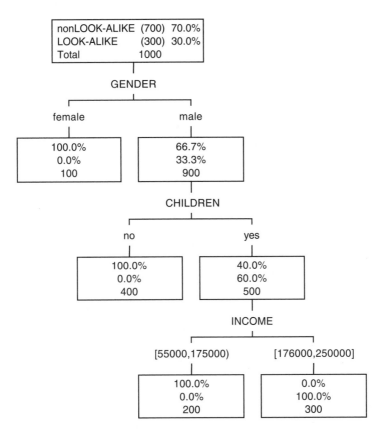

FIGURE 11.7
Look-Alike Tree

The implication is:

1. It is easier to identify an individual who looks like someone with predefined characteristics (for example, gender and children) than someone who behaves in a particular manner (for example, respond to a solicitation).

2. The resultant look-alike rates are biased estimates of target response rates, to the extent the defined look-alike group differs from the target group. Care should be exercised in defining the look-alike group, because it is easy to include individuals inadvertently unwanted.

3. The success of a solicitation based on look-alike profiles, in terms of the actual responses obtained, depends on the disparity of the defined look-alike group and target group, and the tenability of the look-alike assumption.

TABLE 11.5

Gains Chart for Look-Alike Tree Defined by GENDER, CHILDREN and INCOME

Segment*	Size of Segment	Number of Responses	Cumulative Responses	Segment Response Rate	Cumulative Response Rate	Cumulative Lift
#4 - INCOME, [176000,250000) CHILDREN, yes GENDER, male	300	300	300	100.0%	100.0%	333
#2 - CHILDREN, no GENDER, male	400	0	300	0.0%	42.9%	143
#1 - GENDER, female	100	0	300	0.0%	37.5%	125
#3 - INCOME [55000,175000) CHILDREN, yes GENDER, male	200	0	300	0.0%	30.0%	100
	1,000	300		30.0%		

*Segments are ranked by look-alike rates.

11.8 Summary

Database marketers typically attempt to improve the effectiveness of their campaigns by targeting their best customers. However, targeting only the best customers (responders) based on their characteristics is sure to fail. The descriptive profile, which represents responders without reference to non-responders, is a nonpredictive profile. Therefore, targeting with a descriptive profile does not provide any assurance that the best customers will respond to a new campaign. The value of the descriptive profile lies in its definition of the target group's salient characteristics, which are used to develop an effective marketing strategy.

I contrasted descriptive and predictive profiles. A predictive profile describes responders with regard to responsiveness; that is, in terms of variables that discriminate between responders and nonresponders. The predictive profile is used for finding responders to a new campaign, after which the descriptive profile is used to effectively communicate with those customers.

Then, I introduced the tree analysis method of developing a set of complex and interesting predictive profiles. With an illustration, I presented the gains chart as the standard report of the predictive power of the tree-based predictive profiles, for example, the expected response rates upon implementing the profiles in a solicitation.

Last, I expanded the tree-based approach of predictive profiling to Look-Alike profiling, a reliable method when actual response information is not available. A Look-Alike profile is a predictive profile based on a group of individuals who look like the individuals in a target group, thus serving as surrogate responders. I addressed a warning that the look-alike rates are biased estimates of target response rates because the Look-Alike profiles are about surrogate responders, not actual responders.

12

Assessment of Database Marketing Models[1]

Database marketers use the decile analysis to assess their models in terms of classification or prediction accuracy. The uninformed database marketers do not know that additional information from the decile analysis can be extracted to supplement the model assessment. The purpose of this chapter is to present two additional concepts of model assessment — precision and separability — and to illustrate these concepts by further use of the decile analysis.

I begin the discussion with the traditional concepts of accuracy for response and profit models, and illustrate their basic measures of accuracy. Then, I introduce the accuracy measure used in database marketing, known as Cum Lift. The discussion of Cum Lift is set in the context of a decile analysis, which is the usual approach database marketers use to evaluate the performance of response and profit models. I provide a step-by-step procedure for conducting a decile analysis with an illustration.

I continue with the illustration to present the new concepts, precision and separability. Last, I provide guidelines for using all three measures in assessing database marketing models.

12.1 Accuracy for Response Model

How well does a response model correctly classify individuals as responder and nonresponder? The traditional measure of accuracy is the Proportion of Total Correct Classifications (PTCC), which can be calculated from a simple cross-tabulation.

Consider the classification results in Table 12.1 of a response model based on the validation sample consisting of 100 individuals with a 15% response rate. The 'Total' column indicates there are 85 actual nonresponders and 15 actual responders in the sample. The 'Total' row indicates the model predicts 76 nonresponders and 24 responders. The model correctly classifies 74 nonresponders and 13 responders. Accordingly, the PTCC is (74+13)/100 = 87%.

---

[1] This chapter is based on an article with the same title in *Journal of Targeting, Measurement and Analysis for Marketing*, 7, 3, 1998. Used with permission.

TABLE 12.1

Classification Results of Response Model

		Predicted		
		Nonresponder	Responder	Total
Actual	Nonresponder	74	11	85
	Responder	2	13	15
Total		76	24	100

Although PTCC is frequently used, it may not be appropriate for the given situation. For example, if the assessment criterion imposes a penalty for misclassifications,[2] then PTCC must be either modified, or discarded for a more relevant measure.

Database marketers have defined their own measure of accuracy for response models: Cum Lift. They use response models to identify those individuals most likely to respond to a solicitation. They create a solicitation list of the most likely individuals to obtain an advantage over a random selection of individuals. The Cum Lift indicates how many more responses can be expected with a selection based on a model over the expected responses from a random selection (no model). Specifically, the Cum Lift is an index of the expected response rate with a selection based on a model compared with the expected response rate based on a random selection. Before I illustrate the calculation of the Cum Lift, I provide a companion exposition of accuracy for the profit model.

12.2 Accuracy for Profit Model

How well does a profit model correctly predict an individual's profit value? There are several measures of prediction accuracy, all of which are based on the concept of *error*, namely, actual profit minus predicted profit. The Mean Squared Error (MSE) is by far the most popular measure, but it is flawed, thus necessitating three alternative measures. I briefly review the four error measures.

1. MSE is the mean of individual squared errors. It gives greater importance to larger errors and tends to underestimate the predictive accuracy of the model. This point is illustrated below.

2. MPE is the mean of individual percentage errors. It measures the bias in the estimates. Percentage error for an individual is the error divided by the actual profit, multiplied by 100.

---

[2] For example, there is a $2 loss if a responder is classified as a nonresponder; and a $4 loss if a nonresponder is classified as a responder.

3. MAPE is the mean of individual absolute percentage errors. It disregards the sign of the error.

4. MAD is the mean of individual absolute deviations (error). It disregards the sign of the error.

Consider the prediction results in Table 12.2 of the profit model (not shown); the validation sample consists of thirty individuals with a mean profit of $6.76. The first two columns are the *actual* profit and *predicted* profit produced by the profit model. The remaining columns are Error, Squared Error, Percentage Error, Absolute Error, and Absolute Percentage Error. The bottom row MEAN consists of the means, based on the thirty individuals, for the last four columns. The mean values are 23.22, 52.32%, 2.99 and 87.25%, for MSE, MPE, MAD and MAPE, respectively. These measures are only indicators of a good model; "smaller" values tend to correspond to "better" models.

To highlight the sensitivity of MSE to far-out values, I calculate MSE, MPE, MAD and MAPE based on the sample without the large error of 318.58 corresponding to individual #26. The adjusted mean values for MSE, MPE, MAD and MAPE are 13.03, 49.20%, 2.47 and 85.30%, respectively. The sensitivity of MSE is clear as it is dramatically reduced by almost 50%, whereas MPE, MAD and MAPE remain relatively stable.

Except for the occasional need for an individual-level profit accuracy assessment, which requires one of the four error measures, database marketers use their own measure of accuracy — Cum Lift — for profit models. When Cum Lift is used for a profit model — Cum Lift (profit) — it is very similar to the Cum Lift for a response model, except for slight changes in assessment and interpretation. Database marketers use profit models to identify individuals contributing maximum profit from a solicitation and create a solicitation list of those individuals to obtain an advantage over a random selection. The Cum Lift (profit) indicates how much more profit can be expected with a selection based on a model over the expected profit from a random selection with no profit model. Specifically, the Cum Lift (profit) is an index of the expected profit with a selection based on a model compared with the expected profit based on a random selection.

12.3 Decile Analysis and Cum Lift for Response Model

The decile analysis is a tabular display of model performance. I illustrate the construction and interpretation of the response decile analysis in Table 12.3. (The response model, on which the decile analysis is based, is not shown.)

TABLE 12.2

Profit Model: Four Measure of Errors

ID #	PROFIT	Predicted PROFIT	Error	Squared Error	Percentage Error	Absolute Error	Absolute Percentage Error
1	0.60	0.26	0.34	0.12	132.15%	0.34	132.15%
2	1.60	0.26	1.34	1.80	519.08%	1.34	519.08%
3	0.50	0.26	0.24	0.06	93.46%	0.24	93.46%
4	1.60	0.26	1.34	1.80	519.08%	1.34	519.08%
5	0.50	0.26	0.24	0.06	93.46%	0.24	93.46%
6	1.20	0.26	0.94	0.89	364.31%	0.94	364.31%
7	2.00	1.80	0.20	0.04	11.42%	0.20	11.42%
8	1.30	1.80	-0.50	0.25	-27.58%	0.50	27.58%
9	2.50	1.80	0.70	0.50	39.27%	0.70	39.27%
10	2.20	3.33	-1.13	1.28	-33.97%	1.13	33.97%
11	2.40	3.33	-0.93	0.87	-27.96%	0.93	27.96%
12	1.20	3.33	-2.13	4.54	-63.98%	2.13	63.98%
13	3.50	4.87	-1.37	1.87	-28.10%	1.37	28.10%
14	4.10	4.87	-0.77	0.59	-15.78%	0.77	15.78%
15	5.10	4.87	0.23	0.05	4.76%	0.23	4.76%
16	5.70	6.40	-0.70	0.50	-11.00%	0.70	11.00%
17	3.40	7.94	-4.54	20.62	-57.19%	4.54	57.19%
18	9.70	7.94	1.76	3.09	22.14%	1.76	22.14%
19	8.60	7.94	0.66	0.43	8.29%	0.66	8.29%
20	4.00	9.48	-5.48	30.01	-57.80%	5.48	57.80%
21	5.50	9.48	-3.98	15.82	-41.97%	3.98	41.97%
22	10.50	9.48	1.02	1.04	10.78%	1.02	10.78%
23	17.50	11.01	6.49	42.06	58.88%	6.49	58.88%
24	13.40	11.01	2.39	5.69	21.66%	2.39	21.66%
25	4.50	11.01	-6.51	42.44	-59.14%	6.51	59.14%
26	30.40	12.55	17.85	318.58	142.21%	17.85	142.21%

ID #	PROFIT	Predicted PROFIT	Error	Squared Error	Percentage Error	Absolute Error	Absolute Percentage Error
27	12.40	15.62	-3.22	10.40	-20.64%	3.22	20.64%
28	13.40	17.16	-3.76	14.14	-21.92%	3.76	21.92%
29	26.20	17.16	9.04	81.71	52.67%	9.04	52.67%
30	7.40	17.16	-9.76	95.28	-56.88%	9.76	56.88%
			MEAN	23.22	52.32%	2.99	87.25%
			MEAN without ID# 26	13.03	49.20%	2.47	85.30%

TABLE 12.3

Response Decile Analysis

Decile	Number of Individuals	Number of Responders	Decile Response Rate	Cumulative Response Rate	Cumulative Lift
Top	7,410	911	12.3%	12.3%	294
2	7,410	544	7.3%	9.8%	235
3	7,410	437	5.9%	8.5%	203
4	7,410	322	4.3%	7.5%	178
5	7,410	258	3.5%	6.7%	159
6	7,410	188	2.5%	6.0%	143
7	7,410	130	1.8%	5.4%	129
8	7,410	163	2.2%	5.0%	119
9	7,410	124	1.7%	4.6%	110
Bottom	7,410	24	0.3%	4.2%	100
Total	74,100	3,101	4.2%		

1. Score the sample (i.e., calibration or validation file) using the response model under consideration. Every individual receives a model score, Prob_est, the model's estimated probability of response.

2. Rank the scored file, in descending order by Prob_est.

3. Divide the ranked and scored file into ten equal groups. The *Decile* variable is created, which takes on ten ordered "values": top(1), 2, 3, 4, 5, 6, 7, 8, 9, and bottom(10). The "top" decile consists of the best 10% of individuals most-likely to respond; decile 2 consists of the next 10% of individuals most-likely to respond. And so on, for the remaining deciles. Accordingly, decile separates and orders the individuals on an ordinal scale ranging from most- to least-likely to respond.

4. *Number of Individuals* is the number of individuals in each decile; 10% of the total size of the file.

5. *Number of Responses (actual)* is the actual — not predicted — number of responses in each decile. The model identifies 911 actual responders in the top decile. In decile 2, the model identifies 544 actual responders. And so on, for the remaining deciles.

6. *Decile Response Rate* is the actual response rate for each decile group. It is *Number of Responses* divided by *Number of Individuals* for each decile group. For the top decile, the response rate is 12.3% (= 911/7,410). For the second decile, the response rate is 7.3% (= 544/7,410). Similarly for the remaining deciles.

7. *Cumulative Response Rate* for a given depth-of-file (the aggregated or cumulative deciles) is the response rate among the individuals in the cumulative deciles. For example, the cumulative response rate

for the top decile (10% depth-of-file) is 12.3% (= 911/7,410). For the top two deciles (20% depth-of-file), the cumulative response rate is 9.8% = ([911+544]/[7410+7410]). Similarly for the remaining deciles.

8. *Cum Lift* for a given depth-of-file is the *Cumulative Response Rate* divided by the overall response rate of the file, multiplied by 100. It measures how much better one can expect to do with the model than without a model. For example, a Cum Lift of 294 for the top decile means that when soliciting to the top 10% of the file based on the model, one can expect 2.94 times the total number of responders found by randomly soliciting 10%-of-file. The Cum Lift of 235 for the top two deciles means that when soliciting to 20% of the file based on the model, one can expect 2.35 times the total number of responders found by randomly soliciting 20%-of-file without a model. Similarly for the remaining deciles.

Rule: The larger the Cum Lift value, the better the accuracy for a given depth-of-file.

12.4 Decile Analysis and Cum Lift for Profit Model

Calculation of the profit decile analysis is very similar to that of the response decile analysis with "response" and "response rates" replaced by "profit" and "mean profit," respectively. I illustrate the construction and interpretation of the profit decile analysis in Table 12.4

TABLE 12.4

Profit Decile Analysis

Decile	Number of Individuals	Total Profit	Decile Mean Profit	Cumulative Mean Profit	Cumulative Lift
Top	3	$47.00	$15.67	$15.67	232
2	3	$60.30	$20.10	$17.88	264
3	3	$21.90	$7.30	$14.36	212
4	3	$19.40	$6.47	$12.38	183
5	3	$24.00	$8.00	$11.51	170
6	3	$12.70	$4.23	$10.29	152
7	3	$5.80	$1.93	$9.10	135
8	3	$5.80	$1.93	$8.20	121
9	3	$2.70	$0.90	$7.39	109
Bottom	3	$3.30	$1.10	$6.76	100
Total	30	$202.90	$6.76		

1. Score the sample (i.e. calibration or validation file) using the profit model under consideration. Every individual receives a model score, Pred_est, the model's predicted profit.

2. Rank the scored file, in descending order by Pred_est.

3. Divide the ranked and scored file into ten equal groups, producing the *decile* variable. The "top" decile consists of the best 10% of individuals contributing maximum profit; decile 2 consists of the next 10% of individuals contributing maximum profit. And so on, for the remaining deciles. Accordingly, decile separates and orders the individuals on an ordinal scale ranging from maximum to minimum contribution of profit.

4. *Number of Individuals* is the number of individuals in each decile; 10% of the total size of the file.

5. *Total Profit (actual)* is the actual — not predicted — total profit in each decile. The model identifies individuals contributing $47 profit in the top decile. In decile 2, the model identifies individuals contributing $60.30 profit. And so on, for the remaining deciles.

6. *Decile Mean Profit* is the actual mean profit for each decile group. It is *Total Profit* divided by *Number of Individuals* for each decile group. For the top decile, the actual mean profit is $15.67 (= $47/3). For the second decile, the value is $20.10 (= $60.30/3). Similarly for the remaining deciles.

7. *Cumulative Mean Profit* for a given depth-of-file (the aggregated or cumulative deciles) is the mean profit among the individuals in the cumulative deciles. For example, the cumulative mean profit for the top decile (10% depth-of-file) is $15.67 (= $47/3). For the top two deciles (20% depth-of-file), the cumulative response rate is $17.88 = ([$47+$60.30]/[3+3]). Similarly for the remaining deciles.

8. *Cum Lift* for a given depth-of-file is the *Cumulative Mean Profit* divided by the overall profit of the file, multiplied by 100. It measures how much better one can expect to do with the model than without a model. For example, a Cum Lift of 232 for the top decile means that when soliciting to the top 10% of the file based on the model, one can expect 2.32 times the total profit found by randomly soliciting 10%-of-file. The Cum Lift of 264 for top two deciles means that when soliciting to 20%-of-file based on the model, one can expect 2.64 times the total profit found by randomly soliciting 20%-of-file. Similar for the remaining deciles. Note that the non-decreasing profit values throughout the decile suggest that something is "wrong" with this model, e.g., an important predictor variable is not included, or a predictor variable in the model needs to be re-expressed.

Rule: The larger the Cum Lift value, the better the accuracy for a given depth-of-file.

12.5 Precision for Response Model

How close are the predicted probabilities of response to the true probabilities of response? Closeness or precision of the response cannot directly be determined because an individual's true probability is not known — if it were, then a model would not be needed! I recommend and illustrate the method of *smoothing* to provide estimates of the true probabilities. Then, I present the HL index as the measure of precision for response models.

Smoothing is the averaging of values within "neighborhoods." In this application of smoothing, I average actual responses within "decile" neighborhoods formed by the model. Continuing with the response model illustration, the actual response rate for a decile, Column 4 in Table 12.3 is the estimate of true probability of response for the group of individuals in that decile.

Next, I calculate the mean predicted probabilities of response based on the response model scores (Prob_est) among the individuals in each decile. I insert these predicted means in Column 4 in Table 12.5. Also, I insert the actual response rate (Column 4 in Table 12.3) in Column 3 in Table 12.5. I can now assess response model precision.[3]

Comparing Columns 3 and 4 (in Table 12.5) is informative. I see that for the top decile, the model underestimates the probability of response: 12.3% actual vs. 9.6% predicted. Similarly, the model underestimates for deciles 2 through 4. Decile 5 is perfect. Going down deciles 6 through bottom, it becomes clear that the model is overestimating. This type of evaluation for precision is perhaps too subjective. An objective summary measure of precision is needed.

I present the HL index[4] as the measure of precision. The calculations for the HL index are illustrated in Table 12.5 for the response model illustration:

1. Columns 1, 2 and 3 are available from the response decile analysis.
2. Calculate the mean predicted probability of response for each decile from the model scores, Prob_est (Column 4).
3. Calculate Column 5: take the difference between Column 3 and Column 4. Square the results. Then, multiply by Column 1.
4. Column 6: Column 4 times the quantity one minus Column 4.
5. Column 7: Column 5 divided by Column 6.
6. HL index: sum of the ten elements of Column 7.

Rule: The smaller the HL index value, the better the precision.

---

[3] This assessment is considered at a 10% level of smooth. A ventile-level analysis with the scored and ranked file divided into twenty groups, would provide an assessment of model precision at a 5% level of smooth. There is no agreement on a reliable level of smooth among statisticians.
[4] This is the Hosmer-Lemeshow goodness-of-fit measure.

TABLE 12.5

Response Model: HL and CV Indices

Decile	Column 1 Number of Individuals	Column 2 Number of Responders	Column 3 Decile Response Rate (Actual)	Column 4 Prob_Est (Predicted)	Column 5 Square of (Column 3 minus Column 4) Times Column 1	Column 6 Column 4 Times (1−Column 4)	Column 7 Column 5 Divided by Column 6
Top	7,410	911	12.3%	9.6%	5.40	0.086	62.25
2	7,410	544	7.3%	4.7%	5.01	0.044	111.83
3	7,410	437	5.9%	4.0%	2.68	0.038	69.66
4	7,410	322	4.3%	3.7%	0.27	0.035	7.49
5	7,410	258	3.5%	3.5%	0.00	0.033	0.00
6	7,410	188	2.5%	3.4%	0.60	0.032	18.27
7	7,410	130	1.8%	3.3%	1.67	0.031	52.25
8	7,410	163	2.2%	3.2%	0.74	0.031	23.92
9	7,410	124	1.7%	3.1%	1.45	0.030	48.35
Bottom	7,410	24	0.3%	3.1%	5.81	0.030	193.40
Total	74,100	3,101	4.2%				587.40
	Separability CV		80.23			Precision HL	587.40

12.6 Precision for Profit Model

How close are the predicted profits to the true profits? Just as in the discussion of precision for the response model, closeness cannot directly be determined because an individual's true profit is not known, and I recommend and illustrate the method of smoothing to provide estimates of the true profit values. Then, I present the SWMAD index as the measure of precision for a profit model.

To obtain the estimates of the true profit values, I average actual profit within "decile" neighborhoods formed by the profit model. Continuing with the profit model illustration, the mean actual profit for a decile, Column 4 in Table 12.4, is the estimate of true mean profit for the group of individuals in that decile. Next, I calculate the mean predicted profit based on the model scores (Pred_est) among the individuals in each decile. I insert these predicted means in Column 2 in Table 12.6. Also, I insert the mean actual profit (Column 4 in Table 12.4) in Column 1 in Table 12.6. I can now assess model precision.

Comparing Columns 1 and 2 (in Table 12.6) is informative. The ranking of the deciles based on the mean actual profit values is not strictly descending — not a desirable indicator of a good model. The mean profit values for the top and 2nd deciles are reversed, $15.67 and $20.10, respectively. Moreover, third largest decile mean profit value ($8.00) is in the 5th decile. This type of evaluation is interesting, but a quantitative measure for this "out of order" ranking is preferred.

12.6.1 Construction of SWMAD

I present the measure SWMAD for the precision of a profit model: a weighted mean of the absolute deviation between smooth decile actual and predicted profit values; the weights reflect discordance between the rankings of the smooth decile actual and predicted values. The steps of the calculation of SWMAD for the profit model illustration are below and reported in Table 12.6:

1. Column 1 is available from the profit decile analysis.
2. Calculate the mean predicted profit for each decile from the model scores, Pred_est (Column 2).
3. Calculate Column 3: take the absolute difference between Column 1 and Column 2.
4. Column 4: rank the deciles based on actual profit (Column 1); assign the lowest rank value to the highest decile mean actual profit value. Tied ranks are assigned the mean of the corresponding ranks.

TABLE 12.6

Profit Model: SWMAD and CV Indices

Decile	Column 1 Decile Mean Profit (Actual)	Column 2 Decile Mean Pred_Est (Predicted Profit)	Column 3 Absolute Error	Column 4 Rank of Decile Actual Profit	Column 5 Rank of Decile Predicted Profit	Column 6 Absolute Difference Between Ranks	Column 7 Wt	Column 8 Weighted Error
Top	$15.67	$17.16	$1.49	2	1	1.0	1.10	1.64
2	$20.10	$13.06	$7.04	1	2	1.0	1.10	7.74
3	$7.30	$10.50	$3.20	4	3	1.0	1.10	3.52
4	$6.47	$8.97	$2.50	5	4	1.0	1.10	2.75
5	$8.00	$7.43	$0.57	3	5	2.0	1.20	0.68
6	$4.23	$4.87	$0.63	6	6	0.0	1.00	0.63
7	$1.93	$3.33	$1.40	7.5	7	0.5	1.05	1.47
8	$1.93	$1.80	$0.14	7.5	8	0.5	1.05	0.15
9	$0.90	$0.26	$0.64	10	9.5	0.5	1.05	0.67
Bottom	$1.10	$0.26	$0.84	9	9.5	0.5	1.05	0.88
Separability CV	95.86					SUM	10.8	20.15
						Precision	SWMAD	1.87

5. Column 5: rank the deciles based on predicted profit (Column 2); assign the lowest rank value to the highest decile mean predicted profit value. Tied ranks are assigned the mean of the corresponding ranks.

6. Column 6: take the absolute difference between Column 4 and Column 5.

7. Column 7: the weight variable (Wt) is defined as Column 6 divided by 10 plus 1.

8. Column 8: Column 3 times Column 7.

9. Calculate SUMWGT: the sum of the ten values of Column 7.

10. Calculate SUMWDEV: the sum of the ten values of Column 8.

11. Calculate SWMAD: SUMWDEV/SUMWGT.

Rule: The smaller the SWMAD value, the better the precision.

12.7 Separability for Response and Profit Models

How different are the individuals across the deciles in terms of likelihood to respond or contribution to profit? Is there a real variation or separation of individuals as identified by the model? I can measure the variability across the decile groups by calculating the traditional coefficient of variation (CV) among the decile estimates of true probability of response for the response model, and among the decile estimates of true profit for the profit model.

I illustrate the calculation of CV with the response and profit model illustrations. CV (response) is the standard deviation of the ten smooth values of Column 3 in Table 12.5, divided by the mean of the ten smooth values, multiplied by 100. CV (response) is 80.23 in Table 12.5. CV (profit) is the standard deviation of the ten smooth values of Column 1 in Table 12.6, divided by the mean ten smooth values, multiplied by 100. CV (profit) is 95.86 in Table 12.6.

Rule: The larger the CV value, the better the separability.

12.8 Guidelines for Using Cum Lift, HL/SWMAD and CV

The following are guidelines for selecting the best database model based on the three assessment measures Cum Lift, HL/SWMAD and CV:

1. In general, a good model has a large HL/SWMAD and CV values.

2. If maximizing response rate/mean profit is not the objective of the model, then the best model is among those with the smallest HL/SWMAD values and largest CV values. Because small HL/SWMAD values do not necessarily correspond with large CV values, the data analyst must decide on the best balance of small HL/SWMAD and large CV values for declaring the best model.

3. If maximizing response rate/mean profit is the objective, then the best model has the largest Cum Lift. If there are several models with comparable largest-Cum Lift values, then the model with the "best" HL/SWMAD-CV combination is declared the best model.

4. If decile-level response/profit prediction is the objective of the model, then the best model has the smallest HL/SWMAD value. If there are several models with comparable smallest HL/SWMAD values, then the model with the largest CV value is declared the best.

5. The measure of separability CV itself has no practical value. A model that is selected solely on the largest CV value will not necessarily have good accuracy or precision. Separability should be used in conjunction with the other two measures of model assessment as discussed above.

12.9 Summary

The traditional measures of model accuracy are a proportion of total correct classification and mean square error or a variant of mean error, for response and profit models, respectively. These measures have limited value in database marketing. Database marketers have their own measure of model accuracy — Cum Lift — which takes into account the way they implement the model. They use a model to identify individuals most likely to respond or contribute profit and create a solicitation list of those individuals to obtain advantage over a random selection. The Cum Lift is an index of the expected response/profit with a selection based on a model compared with the expected response/profit with a random selection (no model). A maxim of Cum Lift is, the larger the Cum Lift value, the better the accuracy.

I discussed the Cum Lift by illustrating the construction of the decile analysis for both response and profit models. Then, using the decile analysis as a backdrop, I presented two additional measures of model assessment — HL/SWMAD for response/profit precision, and the traditional coefficient of variation CV for separability for response and profit models. Because the true response/profit values are unknown, I estimated the true values by smoothing at the decile-level. With these estimates, I illustrated the calculations for HL and SWMAD. The HL/SWMAD rule: the smaller the HL/SWMAD values, the better the precision.

Separability addresses the question of how different the individuals are in terms of likelihood to respond or contribution to profit across the deciles. I used traditional coefficient of variation CV as the measure of separability among the estimates of true response/profit values. Thus a maxim of CV is that the larger the CV value, the better the separability.

Last, I provided guidelines for using all three measures together in selecting the best model.

13

Bootstrapping in Database Marketing: A New Approach for Validating Models[1]

Traditional validation of a database marketing model uses a hold-out sample consisting of individuals who are not part of the sample used in building the model itself. The validation results are probably biased and incomplete unless a resampling method is used. This chapter points to the weaknesses of the traditional validation, and then presents a boostrap approach for validating response and profit models, as well as measuring the efficiency of the models.

13.1 Traditional Model Validation

The data analyst's first step in building a database marketing model is to randomly split the original data file into two mutually exclusive parts: a calibration sample for developing the model, and a validation or hold-out sample for assessing the reliability of the model. If the analyst is lucky to split the file to yield a hold-out sample with *favorable* characteristics, then a better-than-true biased validation is obtained. If unlucky, and the sample has *unfavorable* characteristics, then a worse-than-true biased validation is obtained. Lucky or not, or even if the validation sample is a true reflection of the population under study, a single sample cannot provide a measure of variability that would otherwise allow the analyst to assert a level of confidence about the validation.

In sum, the traditional single-sample validation provides neither assurance that the results are not biased, nor any measure of confidence in the results. These points are made clear with an illustration using a response model; but all results and implications equally apply to profit models.

---

[1] This chapter is based on an article with the same title in *Journal of Targeting, Measurement and Analysis for Marketing*, 6, 2, 1997. Used with permission.

13.2 Illustration

As database marketers use the Cum Lift measure from a decile analysis to assess the goodness of a model, the validation of the model[2] consists of comparing the Cum Lifts from the calibration and hold-out decile analyses based on the model. It is expected that shrinkage in the Cum Lifts occurs: Cum Lifts from the hold-out sample are typically smaller (less optimitic) than those from the calibration sample, from which they were originally born. The Cum Lifts on a fresh hold-out sample, which does not contain the calibration idiosyncrasies, provide a more realistic assessment of the quality of the model. The calibration Cum Lifts inherently capitalize on the idiosyncrasies of the calibration sample due to the modeling process that favors large Cum Lifts. If both the Cum Lift shrinkage and the Cum Lift values themselves are acceptable, then the model is considered successfully validated and ready to use; otherwise, the model is reworked until successfully validated.

Consider a response model (RM) that produces the validation decile analysis in Table 13.1 based on a sample of 181,100 customers with an overall response rate 0.26%. (Recall from Chapter 12 that the Cum Lift is a measure of predictive power: it indicates the expected gain from a solicitation implemented *with* a model over a solicitation implemented *without* a model.) The Cum Lift for the top decile is 186; this indicates that when soliciting to the top decile — the top 10% of the customer file identified by the RM model — there is an expected 1.86 times the number of responders found by randomly soliciting 10% of the file (without a model). Similarly for the second

TABLE 13.1

Response Decile Analysis

Decile	Number of Individuals	Number of Responders	Decile Response Rate	Cumulative Response Rate	Cum Lift
Top	18,110	88	0.49%	0.49%	186
2	18,110	58	0.32%	0.40%	154
3	18,110	50	0.28%	0.36%	138
4	18,110	63	0.35%	0.36%	137
5	18,110	44	0.24%	0.33%	128
6	18,110	48	0.27%	0.32%	123
7	18,110	39	0.22%	0.31%	118
8	18,110	34	0.19%	0.29%	112
9	18,110	23	0.13%	0.27%	105
Bottom	18,110	27	0.15%	0.26%	100
Total	181,100	474	0.26%		

---

[2] Validation of any response or profit model built from any modeling technique (e.g., discriminant analysis, logistic regression, neural network, genetic algorithms, or CHAID).

TABLE 13.2

Cum Lifts for Three Validations

Decile	First Sample	Second Sample	Third Sample
Top	186	197	182
2	154	153	148
3	138	136	129
4	137	129	129
5	128	122	122
6	123	118	119
7	118	114	115
8	112	109	110
9	105	104	105
Bottom	100	100	100

decile, the Cum Lift 154 indicates that when soliciting to the top-two deciles — the top 20% of the customer file based on the RM model — there is an expected 1.54 times the number of responders found by randomly soliciting 20% of the file.

As luck would have it, the data analyst finds two additional samples on which two additional decile analysis validations are performed. Not surprising, the Cum Lifts for a given decile across the three validations are somewhat different. The reason for this is the *expected* sample-to-sample variation, attributed to chance. There is a large variation in the top decile (range is 15 = 197 – 182), and a small variation in decile 2 (range is 6 = 154 – 148). These results in Table 13.2 raise obvious questions.

13.3 Three Questions

With many decile analysis validations, the expected sample-to-sample variation within each decile points to the uncertainty of the Cum Lift estimates. If there is an observed large variation for a given decile, there is less confidence in the Cum Lift for that decile; if there is an observed small variation, there is more confidence in the Cum Lift. Thus, the following questions:

1. With many decile analysis validations, how can an *average* Cum Lift (for a decile) be defined to serve as an *honest* estimate of the Cum Lift? Additionally, how many validations are needed?

2. With many decile analysis validations, how can the *variability* of an honest estimate of Cum Lift be assessed? That is, how can the standard error (a measure of precision of an estimate) of honest estimate of the Cum Lift be calculated?

3. With only a single validation dataset, can an honest Cum Lift estimate and its standard error be calculated?

The answers to these questions and more lie in the bootstrap methodology.

13.4 The Bootstrap

The bootstrap is a computer-intensive approach to statistical inference. [1] It
is the most popular resampling method, using the computer to extensively
resample the sample at hand.[3] [2] By random selection with replacement
from the sample, some individuals occur more than once in a *bootstrap*
sample, and some individuals occur not at all. Each same-size bootstrap
sample will be slightly different from one another. This variation makes it
possible to induce an empirical sampling distribution[4] of the desired statistic,
from which estimates of bias and variability are determined.

The bootstrap is a flexible technique for assessing the accuracy[5] of any
statistic. For well-known statistics, such as the mean, the standard deviation,
regression coefficients and R-squared, the bootstrap provides an alternative
to traditional parametric methods. For statistics with unknown properties,
such as the median and Cum Lift, traditional parametric methods do not
exist; thus, the bootstrap provides a viable alternative over the inappropriate
use of traditional methods, which yield questionable results.

The bootstrap falls also into the class of nonparametric procedures. It does
not rely on unrealistic parametric assumptions. Consider testing the significance of a variable[6] in a regression model built using ordinary least-squares
estimation. Say the error terms are not normally distributed, a clear violation
of the least-squares assumptions. [3] The significance testing may yield inaccurate results due to the model assumption not being met. In this situation,
the bootstrap is a feasible approach in determining the significance of the
coefficient without concern of any assumptions. As a nonparametric method,
the bootstrap does not rely on theoretical derivations required in traditional
parametric methods. I review the well-known parametric approach to the
construction of confidence intervals to demonstrate the utility of the bootstrap as an alternative technique.

---

[3] Other resampling methods include: the jackknife, infinitesimal jackknife, delta method, influence function method, and random subsampling.
[4] A sampling distribution can be considered as the frequency curve of a sample statistic from an
infinite number of samples.
[5] Accuracy includes bias, variance, and error.
[6] That is, is the coefficient equal to zero?

13.4.1 Traditional Construction of Confidence Intervals

Consider the parametric construction of a confidence interval for the population mean. I draw a random sample A of five numbers from a population. Sample A consists of (23, 4, 46, 1, 29). The sample mean is 20.60, the sample median is 23, and the sample standard deviation is 18.58.

The parametric method is based on the central limit theorem, which states that the theoretical sampling distribution of the sample mean is normally distributed with an analytically defined standard error. [4] Thus, the $100(1 - a)\%$ confidence interval (CI) for the mean is:

$$\text{Sample Mean value} +/- |Z_{a/2}| * \text{Standard Error}$$

where

Sample Mean value is simply the mean of the five numbers in the sample.

$|Z_{a/2}|$ is the value from the standard normal distribution for a $100(1 - a)\%$ CI. The $|Z_{a/2}|$ values are 1.96, 1.64 and 1.28 for 95%, 90% and 80% CIs, respectively.

Standard Error (SE) of the sample mean has the analytic formula: SE = the sample standard deviation divided by the square root of the sample size.

An often used term is the *margin of error* defined as $|Z_{a/2}| * \text{SE}$.

For sample A, SE equals 8.31, and the 95% CI for the population mean is between 4.31 (= (20.60 – 1.96*8.31) and 36.89 (= (20.60 + 1.96*8.31). This is commonly, although not quite exactly, stated as: there is a 95% confidence that the population mean lies between 4.31 and 36.89. The statistically correct statement for the 95% CI for the unknown population mean is: if one repeatedly calculates such intervals from, say, 100 independent random samples, 95% of the constructed intervals would contain the true population mean. Do not be fooled: once the confidence interval is constructed, the true mean is either in the interval, or not in the interval. Thus, the 95% confidence refers to the procedure for constructing the interval, not the observed interval itself.

The above parametric approach for the construction of confidence intervals for statistics, such as the median or the Cum Lift, does not exist because the theoretical sampling distributions (which provides the standard errors) of the median and Cum Lift are not known. If confidence intervals for median or Cum Lift are desired, then approaches that rely on a resampling methodology like the bootstrap can be used.

13.5 How to Bootstrap[7]

The key assumption of the bootstrap is that the sample is the best estimate[8] of the unknown population. Treating the sample as the population, the analyst repeatedly draws same-size random samples with replacement from the original sample. The analyst estimates the desired statistic's sampling distribution from the many bootstrap samples, and is able to calculate a bias-reduced bootstrap estimate of the statistic, and a bootstrap estimate of the SE of the statistic.

The bootstrap procedure consists of ten simple steps.

1. State desired statistic, say, Y.
2. Treat sample as population.
3. Calculate Y on the sample/population; call it SAM_EST.
4. Draw a bootstrap sample from the population, i.e., a random selection with replacement of size n, the size of the original sample.
5. Calculate Y on the bootstrap sample to produce a pseudo-value; call it BS_1.
6. Repeat steps 4 and 5 "m" times.[9]
7. From steps 1 to 6 there are: $BS_1, BS_2, ..., BS_m$.
8. Calculate the bootstrap estimate of the statistic:[10]
   $BS_{est}(Y) = 2*SAM\_EST - mean(BS_i)$.
9. Calculate the bootstrap estimate of the standard error of the statistic:
   $SE_{BS}(Y)$ = standard deviation of (BS_i).
10. The 95% bootstrap confidence interval is
    $$BS_{est}(Y) +/- |Z_{0.025}| * SE_{BS}(Y).$$

13.5.1 Simple Illustration

Consider a simple illustration. I have a sample B from a population (no reason to assume it's normally distributed), which produced the following eleven values:

Sample B: 0.1   0.1   0.1   0.4   0.5   1.0   1.1   1.3   1.9   1.9   4.7

---

[7] This bootstrap method is the normal approximation. Others are Percentile, B-C Percentile and Percentile-t.

[8] Actually, the sample distribution function is the nonparametric maximum likelihood estimate of the population distribution function.

[9] Studies show the precision of the bootstrap does not significantly increase for m > 250.

[10] This calculation arguably insures a bias-reduced estimate. Some analysts question the use of the bias correction. I feel that this calculation adds precision to the decile analysis validation when conducting small solicitations, and has no noticeable effect on large solicitations.

I want a 95% confidence interval for the population standard deviation. If I knew the population was normal, I would use the parametric chi-square test, and obtain the confidence interval:

$$0.93 < \text{population standard deviation} < 2.35.$$

I apply the bootstrap procedure on sample B:

1. The desired statistic is the standard deviation, StD.
2. Treat sample B as the population.
3. I calculate StD on the original sample/population. SAM_EST = 1.3435.
4. I randomly select 11 observations with replacement from the population. This is the first bootstrap sample.
5. I calculate StD on this bootstrap sample to obtain a pseudo-value, $BS_1 = 1.3478$.
6. I repeat steps 4 and 5 an additional 99 times.
7. I have BS_1, BS_2, ..., BS_{100} in Table 13.3.
8. I calculate the bootstrap estimate of StD:
   $BS_{est}(StD) = 2*SAM\_EST - mean(BS_i) = 2*1.3435 - 1.2034 = 1.483$
9. I calculate the bootstrap estimate of the standard error of StD:
   $SE_{BS}(StD) = $ standard deviation $(BS_i) = 0.5008$.
10. The bootstrap 95% confidence interval for the population standard deviation: $0.50 < \text{population standard deviation} < 2.47$.

As you may have suspected, the sample was drawn from a normal population (NP). Thus, it is instructive to compare the performance of the bootstrap with the theoretically correct parametric chi-square test. The bootstrap (BS) confidence interval is somewhat wider than the chi-square/NP interval in Figure 13.1. The BS confidence interval covers values between .50 and 2.47, which also includes the values within the NP confidence interval (0.93, 2.35).

TABLE 13.3

100 Bootstrapped StDs

1.3476	0.6345	1.7188	...	...	...	...	1.4212	1.6712	1.0758
...			...	...	...	...			...
...			...	...	...	...			...
...			...	...	...	...			...
...			...	...	...	...			...
...			...	...	...	...			...
...			...	...	...	...			...
...			...	...	...	...			...
...			...	...	...	...			...
1.3666	1.5388	1.4211	...	...	...	...	1.4467	1.9938	0.5793

NP   0.93xxxxxxxxxxxxxxx2.35
BS   .50xxxxxxxxxxxxxxxxxxxx2.47

FIGURE 13.1
Bootstrap vs. Normal Estimates

These comparative performances results are typical. Performance studies indicate that the bootstrap methodology provides results that are consistent with the outcomes of the parametric techniques. Thus, the bootstrap is a reliable approach to inferential statistics in most situations.

Note that the bootstrap estimate offers a more honest[11] and bias-reduced[12] point estimate of the standard deviation. The original sample estimate is 1.3435, and the bootstrap estimate is 1.483. There is a 10.4% (1.483/1.3435) bias-reduction in the estimate.

13.6 Bootstrap Decile Analysis Validation

Continuing with the RM model illustration, I execute the ten-step bootstrap procedure to perform a bootstrap decile analysis validation. I use 50 bootstrap samples,[13] each of size equal to the original sample size of 181,100. The bootstrap Cum Lift for the top decile is 183 and has a bootstrap standard error of 10. Accordingly, for the top decile, the 95% bootstrap confidence interval is 163 to 203. The second decile has a bootstrap Cum Lift of 151, and a bootstrap 95% confidence interval is between 137 and 165. Similar readings in Table 13.4 can be made for the other deciles. Specifically, this bootstrap validation indicates that the expected Cum Lift is 135 using the RM model to select the top 30% of the most responsive individuals from a randomly drawn sample of size 181,100 from the target population or database. Moreover, the Cum Lift is expected to lie between 127 and 143 with 95% confidence. Similar statements can be made for other depths-of-file from the bootstrap validation in Table 13.4.

Bootstrap estimates and confidence intervals for decile *response rates* can be easily obtained from the bootstrap estimates and confidence intervals for decile Cum Lifts. The conversion formula is: bootstrap decile response rate equals bootstrap decile Cum Lift divided by 100, then multiplied by overall response rate. For example, for third decile the bootstrap response rate is 0.351% (= (135/100)*0.26%); the lower and upper confidence interval end points are 0.330% (= (127/100)*0.26%) and 0.372% (= (143/100)*0.26%), respectively.

---

[11] Due to the many samples used in the calculation.
[12] Attributable, in part, to sample size.
[13] I experience high precision in bootstrap decile validation with just 50 bootstrap samples.

TABLE 13.4

Bootstrap Response Decile Validation
(Bootstrap Sample Size n = 181,000)

Decile	Bootstrap Cum Lift	Bootstrap SE	95% Bootstrap CI
Top	183	10	(163, 203)
2	151	7	(137, 165)
3	135	4	(127, 143)
4	133	3	(127, 139)
5	125	2	(121, 129)
6	121	1	(119, 123)
7	115	1	(113, 117)
8	110	1	(108, 112)
9	105	1	(103, 107)
Bottom	100	0	(100, 100)

13.7 Another Question

Quantifying the predictive certainty, that is, constructing prediction confidence intervals, is likely to be more informative to data analysts and their management than obtaining a *point estimate* alone. A single calculated value of a statistic such as Cum Lift can be regarded as a point estimate that provides a best guess of the true value of the statistic. However, there is an obvious need to quantify the certainty associated with such a point estimate. The decision maker wants the margin of error of the estimate. Plus/minus what value should be added/substracted to the estimated value to yield an interval, for which there is a reasonable confidence that the true (Cum Lift) value lies? If the confidence interval (equivalently, the standard error or margin of error) is too large for the business objective at hand, what can be done?

The answer rests on the the well-known, fundamental relationship between sample size and confidence interval length: increasing (decreasing) sample size increases (decreases) confidence in the estimate; or equivalantly, increasing (decreasing) sample size decreases (increases) the standard error. [5] The sample size-confidence length relationship can be used to increase confidence in the bootstrap Cum Lift estimates in two ways:

1. If ample additional customers are available, add them to the original validation dataset until the enhanced validation dataset size produces the desired standard error and confidence interval length.

2. Simulate or bootstrap the original validation dataset by increasing the bootstrap sample size until the enhanced bootstrap dataset size produces the desired standard error and confidence interval length.

TABLE 13.5

Bootstrap Response Decile Validation
Bootstrap Sample Size n = 225,000

Decile	Bootstrap Cum Lift	Bootstrap SE	95% Bootstrap CI
Top	185	5	(163, 203)
2	149	3	(137, 165)
3	136	3	(127, 143)
4	133	2	(127, 139)
5	122	1	(121, 129)
6	120	1	(119, 123)
7	116	1	(113 ,117)
8	110	0.5	(108, 112)
9	105	0.5	(103, 107)
Bottom	100	0	(100, 100)

Returning to the RM illustration, I increase the bootstrap sample size to 225,000 from the original sample size of 181,100 to produce a slight decrease in the standard error from 4 to 3 for the cumulative top-three deciles. This simulated validation in Table 13.5 indicates that a Cum Lift of 136 centered in a slightly shorter 95% confidence interval between 130 and 142 is expected when using the RM model to select the top 30% of the most responsive individuals from a randomly selected sample of size 225,000 from the database. Note that the bootstrap Cum Lift estimates also change (in this instance, from 135 to 136), because their calculations are based on new larger samples. Bootstrap Cum Lift estimates rarely show big differences when the enhanced dataset size increases. (Why?) In the next section, I continue the discussion on the sample size–confidence length relationship as it relates to a bootstrap assessment of model implementation performance.

13.8 Bootstrap Assessment of Model Implementation Performance

Statisticians are often asked, "how large a sample do I need to have confidence in my results?" The traditional answer, which is based on parametric theoretical formulations, depends on the statistic in question, such as response rate or mean profit and on some additional required input. These additions include the following:

1. The expected value of the desired statistic.
2. The preselected level of confidence, such as the probability that the decision maker is willing to take by wrongly rejecting the null

hypothesis; this may involve claiming a relationship exists when, in fact, it does not.

3. The preselected level of power to detect a relationship. For example, 1 minus the probability the decision maker is willing to take by wrongly accepting the null hypothesis; this may involve claiming that a relationship does not exist when, in fact, it does.

4. The assurance that a sundry of theoretical assumptions hold.

Regardless of who has built a database model, once it is ready for implementation, the question is essentially the same: "how large a sample do I need to implement a solicitation *based on the model* to obtain a desired performance quantity?" The database answer, which in this case does not require most of the traditional input, depends on one of two performance objectives. Objective one is to maximize the performance quantity for a specific depth-of-file, namely, the Cum Lift. Determining how large the sample should be — actually the smallest sample necessary to obtain Cum Lift value — involves the concepts discussed in the previous section that correspond to the relationship between confidence interval length and sample size. The following procedure answers the sample size question for objective one.

1. For a desired Cum Lift value, identify the decile and its confidence interval containing Cum Lift values closest to the desired value, based on the decile analysis validation at hand. If the corresponding confidence interval length is acceptable, then the size of the validation dataset is the required sample size. Draw a random sample of that size from the database.

2. If the corresponding confidence interval is too large, then increase the validation sample size by adding individuals or bootstrapping a larger sample size until the confidence interval length is acceptable. Draw a random sample of that size from the database.

3. If the corresponding confidence interval is unnecessarily small, which indicates that a smaller sample can be used to save time and cost of data retrieval, then decrease the validation sample size by deleting individuals or bootstrapping a smaller sample size until the confidence interval length is acceptable. Draw a random sample of that size from the database.

Objective two adds a constraint to objective one's performance quantity. Sometimes, it is desirable not only to reach a performance quantity, but also discriminate among the *quality* of individuals who are selected to contribute to the performance quantity. This is particularly worth doing when a solicitation is targeted to an unknown yet believed to be a relatively homogeneous population of finely graded individuals in terms of responsiveness or profitability.

The quality constraint imposes the restriction that the performance value of individuals within a decile is different across the deciles within the preselected depth-of-file range. This is accomplished by a) determining the sample size that produces decile confidence intervals that do not overlap *and* b) insuring an overall confidence level that the individual decile confidence intervals are *jointly* valid; that is, the true Cum Lifts are "in" their confidence intervals. The former condition is accomplished by increasing sample size. The latter condition is accomplished by employing the Bonferroni method, which allows the analyst to assert a confidence statement that multiple confidence intervals are *jointly* valid.

Briefly, the Bonferroni method is as follows: assume the analyst wishes to combine k confidence intervals that individually have confidence levels, $1-a_1$, $1-a_2$, ..., $1-a_k$. She wants to make a joint confidence statement with confidence level $1-a_J$. The Bonferroni method states that the joint confidence level, $1-a_J$, is greater than or equal to $1-a_1 -a_2 ... -a_k$. The joint confidence level is a conservative lower bound on the actual confidence level for a joint confidence statement. The Bonferroni method is conservative in the sense that is provides confidence intervals that have confidence levels larger than the actual level.

I apply the Bonferrroni method for four common individual confidence intervals, 95%, 90%, 85% and 80%.

1. For combining 95% confidence intervals, there is at least 90% confidence that the two true decile Cum Lifts lie between their respective confidence intervals; there is at least 85% confidence that the three true decile Cum Lifts lie between their respective confidence intervals; and there is at least 80% confidence that the four true decile Cum Lifts lie between their respective confidence intervals.

2. For combining 90% confidence intervals, there is at least 80% confidence that the two true decile Cum Lifts lie between their respective confidence intervals; there is at least 70% confidence that the three true decile Cum Lifts lie between their respective confidence intervals; there is at least 60% confidence that the four true decile Cum Lifts lie between their respective confidence intervals.

3. For combining 85% confidence intervals, there is at least 70% confidence that the three two decile Cum Lifts lie between their respective confidence intervals; there is at least 55% confidence that the three true decile Cum Lifts lie between their respective confidence intervals; and there is at least 40% confidence that the four true decile Cum Lifts lie between their respective the confidence intervals.

4. For combining the 80% confidence intervals, there is at least 60% confidence that the three two decile Cum Lifts lie between their

respective confidence intervals; there is at least 40% confidence that the three true decile Cum Lifts lie between their respective the confidence intervals; and there is at least 20% confidence that the four true decile Cum Lifts lie between their respective confidence intervals

The following procedure is used for determining how large the sample should be. Actually, the smallest sample necessary to obtain the desired quantity and quality of performance involves:

1. For a desired Cum Lift value for a preselected depth-of-file range, identify the decile and its confidence interval containing Cum Lift values closest to the desired value based on the decile analysis validation at hand. If the decile confidence intervals within the preselected depth-of-file range do not overlap (or concededly have accetpable minimual overlap), then validation sample size at hand is the required sample size. Draw a random sample of that size from the database.

2. If the decile confidence intervals within the preselected depth-of-file range are too large and/or overlap, then increase the validation sample size by adding individuals or bootstrapping a larger sample size until the decile confidence interval lengths are acceptable and do not overlap. Draw a random sample of that size from the database.

3. If the decile confidence intervals are unnecessarily small and do not overlap, then decrease the validation sample size by deleting individuals or bootstrapping a smaller sample size until the decile confidence interval lengths are acceptably and do not overlap. Draw a random sample of that size from the database.

13.8.1 Illustration

Consider a three-variable response model predicting response from a population with an overall response rate of 4.72%. The decile analysis validation based on a sample size of 22,600 along with the bootstrap estimates are in Table 13.6. The 95% margin of errors ($|Z_{a/2}| * SE_{BS}$) for the top four decile 95% confidence intervals are considered too large for using the model with confidence. Moreover, the decile confidence intervals severely overlap. The top decile has a 95% confidence interval of 160 to 119, with an expected bootstrap response rate of 6.61% (= (140/100)*4.72%; 140 is the bootstrap Cum Lift for the top decile). The top-two decile level has a 95% confidence interval of 113 to 141, with an expected bootstrap response rate of 5.99% (= (127/100)*4.72%; 127 is the bootstrap Cum Lift for the top-two decile levels).

TABLE 13.6
Three-Variable Response Model 95% Bootstrap Decile Validation
Bootstrap Sample Size n = 22,600

Decile	Number of Individuals	Number of Responses	Response Rate (%)	Cum Response Rate (%)	Cum Lift	Bootstrap Cum Lift	95% Margin of Error	95% Lower Bound	95% Upper Bound
Top	2,260	150	6.64	6.64	141	140	20.8	119	160
2nd	2,260	120	5.31	5.97	127	127	13.8	113	141
3rd	2,260	112	4.96	5.64	119	119	11.9	107	131
4th	2,260	99	4.38	5.32	113	113	10.1	103	123
5th	2,260	113	5.00	5.26	111	111	9.5	102	121
6th	2,260	114	5.05	5.22	111	111	8.2	102	119
7th	2,260	94	4.16	5.07	107	107	7.7	100	115
8th	2,260	97	4.29	4.97	105	105	6.9	98	112
9th	2,260	93	4.12	4.88	103	103	6.4	97	110
Bottom	2,260	75	3.32	4.72	100	100	6.3	93	106
Total	22,600	1,067							

TABLE 13.7

Three-Variable Response Model 95% Bootstrap Decile Validation
Bootstrap Sample Size n = 50,000

Decile	Model Cum Lift	Bootstrap Cum Lift	95% Margin of Error	95% Lower Bound	95% Upper Bound
Top	141	140	15.7	124	156
2nd	127	126	11.8	114	138
3rd	119	119	8.0	111	127
4th	113	112	7.2	105	120
5th	111	111	6.6	105	118
6th	111	111	6.0	105	117
7th	107	108	5.6	102	113
8th	105	105	5.1	100	111
9th	103	103	4.8	98	108
Bottom	100	100	4.5	95	104

TABLE 13.8

Three-Variable Response Model 95% Bootstrap Decile Validation
Bootstrap Sample Size n = 75,000

Decile	Model Cum Lift	Bootstrap Cum Lift	95% Margin of Error	95% Lower Bound	95% Upper Bound
Top	141	140	12.1	128	152
2nd	127	127	7.7	119	135
3rd	119	120	5.5	114	125
4th	113	113	4.4	109	117
5th	111	112	4.5	107	116
6th	111	111	4.5	106	116
7th	107	108	4.2	104	112
8th	105	105	3.9	101	109
9th	103	103	3.6	100	107
Bottom	100	100	3.6	96	104

I create a new bootstrap validation sample of size 50,000. The 95% margins of error in Table 13.7 are still unacceptably large and the decile confidence intervals still overlap. I further increase the bootstrap sample size to 75,000 in Table 13.8. There is no noticeable change in 95% margins of error, and the decile confidence intervals overlap.

I recalculate the bootstrap estimates using 80% margins of error on the bootstrap sample size of 75,000. The results in Table 13.9 show that the decile confidence intervals have acceptable lengths and are almost non-overlapping. Unfortunately, the joint confidence levels are quite low: at least 60%, at least 40% and at least 20% for the top-two, top-three and top-four decile levels, respectively.

TABLE 13.9

Three-Variable Response Model 80% Bootstrap Decile Validation
Bootstrap Sample Size n = 75,000

Decile	Model Cum Lift	Bootstrap Cum Lift	80% Margin of Error	80% Lower Bound	80% Upper Bound
Top	141	140	7.90	132	148
2nd	127	127	5.10	122	132
3rd	119	120	3.60	116	123
4th	113	113	2.90	110	116
5th	111	112	3.00	109	115
6th	111	111	3.00	108	114
7th	107	108	2.70	105	110
8th	105	105	2.60	103	108
9th	103	103	2.40	101	106
Bottom	100	100	2.30	98	102

TABLE 13.10

Three-Variable Response Model 95% Bootstrap Decile Validation
Bootstrap Sample Size n = 175,000

Decile	Model Cum Lift	Bootstrap Cum Lift	95% Margin of Error	95% Lower Bound	95% Upper Bound
Top	141	140	7.4	133	147
2nd	127	127	5.1	122	132
3rd	119	120	4.2	115	124
4th	113	113	3.6	109	116
5th	111	111	2.8	108	114
6th	111	111	2.5	108	113
7th	107	108	2.4	105	110
8th	105	105	2.3	103	108
9th	103	103	2.0	101	105
Bottom	100	100	1.9	98	102

After successively increasing the bootstrap sample size, I reach a bootstrap sample size (175,000) which produces acceptable and virtually nonoverlapping 95% confidence intervals in Table 13.10. The top four decile Cum Lift confidence intervals are: (133, 147), (122, 132), (115, 124) and (109, 116), respectively. The joint confidence for the top three deciles and top four deciles are at respectable at-least 85% and at-least 80% levels, respectively. Interesting to note that increasing the bootstrap sample size to 200,000, for either 95% or 90% margins of error, does not produce any noticeable improvement in the quality of performance. (Validations for bootstrap sample size 200,000 are not shown.)

13.9 Bootstrap Assessment of Model Efficiency

The bootstrap approach to decile analysis of database models can provide an assessment of model efficiency. Consider an alternative model A to the the best model B (both models predict the same dependent variable). Model A is said to be *less efficient* than model B if either: 1) model A yields the same results as model B (equal Cum Lift margins of error), where model A sample size is larger than model B sample size; or 2) model A yields worse results than model B (greater Cum Lift margins of error), where sample sizes for models A and B are equal.

The measure of efficiency is reported as the ratio of either quantity: the sample sizes necessary to achieve equal results, or the variablity measures (Cum Lift margin of error) for equal sample size. Efficiency ratio is defined as: model B (quantity) over model A (quantity). Efficiency ratio values less (greater) than 100% indicate that model A is less (more) efficient than model B. In this section, I illustrate how a model with unnecessary predictor variables is less efficient (has larger prediction error variance) than a model with the right number of predictor variables.

Returning to the previous section with the illustration of the three-variable response model (considered the best model B), I create an alternative model (model A) by adding five unnecessary predictor variables to model B. The extra variables include irrelevant variables (not affecting the response variable) and redundant variables (not adding anything to the response variable). Thus, the eight-variable model A can be considered to be an overloaded and noisy model, which should produce unstable predictions with large error variance. The bootstrap decile analysis validation of model A based on a bootstrap sample size of 175,000 are in Table 13.11. To facilitate the discussion, I added the Cum Lift margins of error for model B (from Table 13.10) to Table 13.11. The efficiency ratios for the decile Cum Lifts are reported in the last column of Table 13.11. Note that the decile confidence intervals overlap.

It is clear that model A is less efficient than model B. Efficiency ratios are less than 100%, ranging from a low 86.0% for the top decile to 97.3% for the fourth decile, with one exception for the eighth decile, which has an anomalous ratio of 104.5%. The implication is that model A predictions are less unstable than model B predictions, that is, model A has larger prediction error variance relative to the model B.

The broader implication is that of a warning: review a model with too many variables for justification of each variable's contribution in the model; otherwise, it can be anticipated that the model has unnecessarily large prediction error variance. In other words, the bootstrap methodology should be applied during the model-building stages. A bootstrap decile analysis of model calibration, similar to bootstrap decile analysis of model validation, can be another technique for variable selection, and other assessments of model quality.

TABLE 13.11

Bootstrap Model Efficiency
Bootstrap Sample Size n = 175,000

Decile	Model Cum Lift	Eight-Variable Response Model A				Three-Variable Response Model B	
		Bootstrap Cum Lift	95% Margin of Error	95% Lower Bound	95% Upper Bound	95% Margin of Error	Efficiency Ratio
Top	139	138	8.6	129	146	7.4	86.0%
2nd	128	128	5.3	123	133	5.1	96.2%
3rd	122	122	4.3	117	126	4.2	97.7%
4th	119	119	3.7	115	122	3.6	97.3%
5th	115	115	2.9	112	117	2.8	96.6%
6th	112	112	2.6	109	114	2.5	96.2%
7th	109	109	2.6	107	112	2.4	92.3%
8th	105	105	2.2	103	107	2.3	104.5%
9th	103	103	2.1	101	105	2.0	95.2%
Bottom	100	100	1.9	98	102	1.9	100.0%

13.10 Summary

Traditional validation of database models involves comparing the Cum Lifts from the calibration and hold-out decile analyses based on the model under consideration. If the expected shrinkage (difference in decile Cum Lifts between the two analyses) and the Cum Lifts values themselves are acceptable, then the model is considered successfully validated and ready to use; otherwise, the model is reworked until successfully validated. I illustrated with a response model case study that the single-sample validation provides neither assurance that the results are not biased, nor any measure of confidence in the Cum Lifts.

I proposed the bootstrap — a computer-intensive approach to statistical inference — as a methodology for assessing the bias and confidence in the Cum Lift estimates. I provided a brief introduction to the bootstrap, along with a simple ten-step procedure for bootstrapping any statistic. I illustrated the procedure for a decile validation of the response model in the case study. I compared and contrasted the single-sample and bootstrap decile validations for the case study. It was clear that the bootstrap provides necessary information for a complete validation: the biases and the margins of error of decile Cum Lift estimates.

I addressed the issue of margins of error (confidence levels) that are too large (low) for the business objective at hand. I demonstrated how to use the bootstrap to decrease the margins of error.

Then I continued the discussion on the margin of error to a bootstrap assessment of model implementation performance. I addressed the issue: "how large a sample do I need to implement a solicitation based on the model to obtain a desired performance quantity?" Again, I provided a bootstrap procedure for determining the smallest sample necessary to obtain the desired quantity and quality of performance and illustrated the procedure with a three-variable response model.

Last, I showed how the bootstrap decile analysis can be used an assessment of model efficiency. Continuing with the three-variable response model study, I illustrated the efficiency of the three-variable response model relative to the eight-variable alternative model. The latter model was shown to have more unstable predictions than the former model. The implication is: review a model with too many variables for justification of each variable's contribution in the model; otherwise, it can be anticipated that the model has unnecessarily large prediction error variance. The broader implication is: a bootstrap decile analysis of model calibration, similar to bootstrap decile analysis of model validation, can be another technique for variable selection, and other assessments of model quality.

References

1. Noreen, E.W., *Computer Intensive Methods for Testing Hypotheses*, John Wiley & Sons, New York, 1989.
2. Efron, B., *The Jackknife, the Bootsrap and Other Resampling Plans*, SIAM, Philadelphia, 1982.
3. Draper, N.R. and Smith, H., *Applied Regression Analysis*, John Wiley & Sons, New York, 1966.
4. Neter, J. and Wasserman, W., *Applied Linear Statistical Models*, Irwin, Inc., 1974.
5. Hayes, W.L., *Statistics for the Social Sciences*, Holt, Rinehart and Winston, Austin, TX, 1973.

14

Visualization of Database Models[1]

Visual displays — commonly known as graphs — have been around for a long time, but are currently at their peak of popularity. This is due to the massive amounts of data flowing from the digital environment and the accompanying increase in data visualization software, which provides a better picture of *big data* than ever before. In database marketing, visual displays are commonly used in the exploratory phase of data analysis and model building. An area of untapped potential for visual displays is describing "what the final model is doing" upon implementation of its intended task of predicting. Such displays would increase the confidence in the model builder, while engendering confidence in the marketer, the end-user of the model. In this chapter, I present two graphical methods — the star graphs and profile curves — to visualize the characteristics and performance levels of the individuals, who are predicted by the model.

14.1 Brief History of the Graph

The first visual display had its pictorial debut about 2000 BC when the Egyptians created a real estate map describing data such as property outlines and owner. The Greek Ptolemy created the first world map circa 150 AD using latitudinal and longitudinal lines as coordinates to represent the earth on a flat surface. In the fifteenth century Descartes realized that Ptolemy's geographic map making can serve as a graphical method to identify the relationship between numbers and space, such as patterns. [1] Thus, the common graph was born: a horizontal line (x-axis) and a vertical line (y-axis) intersecting to create a visual space, which occupies numbers defined by an ordered pair of x-y coordinates. The original Descartes graph, which has been embellished with more than 500 years of knowledge and technology, is the genesis of the discipline of *data visualization*, which is experiencing an unprecedented growth due to the advances in microprocessor technology, and plethora of visualization software.

---

[1] This chapter is based on the following: Ratner, B., Profile curves: a method of multivarate comparison of groups, *The DMA Research Council Journal*, 1999. Used with permission.

The scientific community was slow to embrace the Descartes graph, first on a limited basis between the seventeenth century to the mid-eighteenth century, then with much more enthusiasm toward the end of the eighteenth century. [2,3] At the end of the eighteenth century, Playfield initiated work in the area of statistical graphics with the introduction of the bar diagram. Following Playfield's progress with graphical methods, Fourier presented the cumulative frequency polygon, and Quetelet created the frequency polygon and histogram. [4] In 1857, Florence Nightingale — who was a self-educated statistician — invented the pie chart, which she used in her report to the royal commission to force the British army to maintain nursing and medical care to soldiers in the field. [5]

In 1977, Tukey started a revolution of numbers and space with his seminal book *Exploratory Data Analysis* (EDA). [6] Tukey explained, by setting forth in careful and elaborate detail, the unappreciated value of numerical, counting and graphical detective work, performed by simple arithmetic and easy-to-draw pictures. Almost three decades after reinventing the concept of graph-making as a means of encoding numbers for informational and strategical viewing, Tukey's "graphic" off-springs are everywhere. They include the box-and-whisker plot taught in early grade schools, and easily generated computer animated and interactive displays in three-dimensional with 48-bit color used as a staple in business presentations. Tukey, who has been called the "Picasso of Statistics," has visibly left his imprint on today's visual display methodology. [6]

In the preceding chapters I have presented model-based graphical methods, which include smooth and fitted graphs and other Tukey-esque displays for identifying structure of data and fitting of models. Geometry-based graphical methods, which show how the dependent variable varies over the pattern of internal-model variables (variables defining the model), is an area that has not enjoyed growth and can benefit from a Tukey-esque innovation. [7] In this chapter I introduce new applications of two underused, existing methods to show "what the final model is doing," that is, to visualize the individuals identified by the model in terms of *variables of interest* — internal-model variables and/or external-model variables (variables not defining the model) — and the individuals' levels of performance. I illustrate the new applications using a response model; but they equally apply to a profit model.

14.2 Star Graph Basics

A table of numbers neatly shows "important facts contained in a jungle of figures." [6] However, more often than not, the table leaves room for further untangling of the numbers. Graphs can help the data analyst out of the "numerical" jungle with a visual display of where she has been and what she has seen. The data analyst can go only so far into the thicket of numbers until the eye-brain connection needs a graph to extract patterns and gather insight from within the table.

A star graph[2] is a visual display of multivariate data, for example, a table of many rows of many variables. [8] It is especially effective for a small table, like the decile analysis table. The basics of star graph construction are as follows:

1. Identify the units of the star graphs: the *j-observations* and the *k-variables*. Consider a set of j observations. Each observation, which corresponds to a star graph, is defined by an array or row of k variables X.

2. There are k equidistant rays emitting from the center of the star, for each observation.

3. The lengths of the rays correspond to the row of X values. The variables are assumed to be measured on relatively similar scales. If not, the data must be transformed to induce comparable scales. A preferred method to accomplish comparable scales is *standardization*, which transforms all the variables to have the same mean value and the same standard deviation value. The mean value used is essentially arbitrary, but must satisfy the constraint that the transformed standardized values are positive. The standard deviation value used is arbitrary, but is typically given a value of 1. The standardized version of X, Z(X), is defined as: $Z(X) = (X - mean(X))/standard\ deviation(X)$. If X assumes all negative values, there is a preliminary step of multiplying X by –1, producing –X. Then, the standardization is performed on –X.

4. The ends of the rays are connected to form a polygon or star, for each observation.

5. A circle is circumscribed around each star. The circumference provides a reference *line*, which aids in interpreting the star. The centers of the star and circle are the same point. The radius of the circle is equal to the length of the largest ray.

6. A star graph typically does not contain labels indicating the X values. If transformations are required, then the transformed values are virtually *meaningless*.

7. The assessment of the *relative differences in the shapes* of the star graphs untangles the numbers and brings out the true insights within the table.

SAS/Graph has a procedure to generate star graphs. I provide the SAS code at the end of the chapter for the illustrations presented.

---

[2] Star graphs are also known as star glyphs.

14.2.1 Illustration

Database marketers use models to identify potential customers. Specifically, the database model provides a way of identifying individuals into ten equal-sized groups (deciles), ranging from the top 10% most-likely to perform (i.e., respond or contribute profit) to the bottom 10% least-likely to perform. The traditional approach of profiling the individuals identified by the model is to calculate the means of variables of interest and assess their values across the deciles.

Consider a database model for predicting RESPONSE to a solicitation. The marketer is interested in "what is the final model doing," that is, what do the individuals look like in the top decile, in the second decile, etc.? How do the individuals differ across the varying levels of performance (deciles) in terms of the usual demographic variables of interest, AGE, INCOME, EDUCATION and GENDER? Answers to these and related questions provide the marketer with strategic marketing intelligence to put together an effective targeted campaign.

The means of the four demographic variables across the RESPONSE model deciles are in Table 14.1. From the tabular display of means by deciles, I conclude the following:

1. AGE: Older individuals are more responsive than younger individuals.

2. INCOME: High-income earners are more responsive than low-income earners.

3. EDUCATION: Individuals with greater education are more responsive than individuals with less education.

4. GENDER: Females are more responsive than males. Note: A mean GENDER of 0 and 1 implies all females and all males, respectively.

TABLE 14.1

Response Decile Analysis: Demographic Means

Decile	AGE (yrs)	INCOME ($000)	EDUCATION (Years of Schooling)	GENDER (1 = Male/ 0 = Female)
Top	63	155	18	0.05
2	51	120	16	0.10
3	49	110	14	0.20
4	46	111	13	0.25
5	42	105	13	0.40
6	41	95	12	0.55
7	39	88	12	0.70
8	37	91	12	0.80
9	25	70	12	1.00
Bottom	25	55	12	1.00

This interpretation is correct, albeit not thorough because it only considers one variable at a time, a topic covered later in the chapter. It does describe the individuals identified by the model in terms of the four variables of interest and responsiveness. However, it does not beneficially stimulate the marketer into strategic thinking for insights. A graph — *an imprint of many insights* — is needed.

14.3 Star Graphs for Single Variables

The first step in constructing a star graph is to identify the units, which serve as the observations and the variables. For decile-based single-variable star graphs, the variables of interest are the j-observations, and the ten deciles (top, 2, 3, ..., bot) are the k-variables. For the RESPONSE model illustration, there are four star graphs in Figures 14.1, one for each demographic variable. Each star has ten rays, which correspond to the ten deciles.

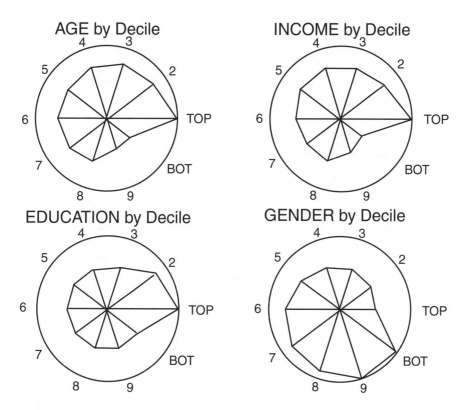

FIGURE 14.1
Star Graphs for AGE, INCOME, EDUCATION, and GENDER.

I interpret the star graphs as follows.

1. For AGE, INCOME and EDUCATION: There is a decreasing trend in the mean values of the variables, as the individuals are successively assigned to the top decile down through the bottom decile. These star graphs visually display: older individuals are more responsive than younger individuals; high income earners are more responsive than low income earners; and individuals with greater education are more responsive than individuals with less education.

2. AGE and INCOME star graphs are virtually identical, except for the ninth decile, which has a slightly protruding vertex. Implication: AGE and INCOME have a similar effect on RESPONSE. Specifically, a standardized unit increase in AGE and in INCOME produce a similar change in RESPONSE.

3. For GENDER: There is an increasing trend in the incidence of males, as the individuals are successively assigned to the top decile down through the bottom decile. Keep in mind GENDER is coded zero for females.

In sum, the star graphs provide a unique visual display of the conclusions in Section 14.2.1. However, they are one-dimensional portraits of each variable's effect on RESPONSE. To obtain a deeper understanding of how the model works as it assigns individuals into the decile analysis table, a full profile of the individuals using all variables *considered jointly* is needed. A multiple-variable star graph provides an unexampled full profile.

14.4 Star Graphs for Many Variables Considered Jointly

As with the single-variable star graph, the first step in constructing the many-variable star graphs is to identify the units. For decile-based many-variable star graphs, the ten deciles are the j-observations, and the variables of interest are the k-variables. For the RESPONSE model illustration, there are ten star graphs, one for each decile, in Figures 14.2. Each decile star has four rays which correspond to the four demographic variables.

I interpret a star graph for an array of variables in a *comparative* context. Because star graphs have no numerical labels, I assess the *shapes* of the stars by observing their *movement* within the reference circle as I go from the top to bottom deciles.

1. *Top decile star:* The rays of AGE, INCOME and EDUCATION touch or nearly touch the circle circumference. These long rays indicate older, more educated individuals with higher income. The short ray

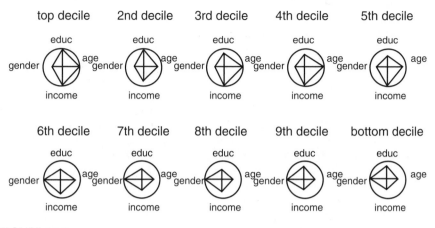

FIGURE 14.2

Star Graphs for Four Demographic Variables Considered Jointly.

of GENDER indicates these individuals are mostly females. The individuals in the top decile comprise the reference group for a comparative analysis of the other decile stars.

2. *2nd decile star:* Individuals in this decile are slightly younger, with less income than individuals in the top decile.

3. *3rd to 5th decile stars:* Individuals in these deciles are substantially less educated than the individuals in the top two deciles. Additionally, the education level decreases as the individuals are assigned to deciles three through five.

4. *6th decile star:* The shape of this star makes a significant departure from the top five decile stars. This star indicates individuals who are mostly males (because the GENDER ray touches the circle), younger, less educated individuals with less income than the individuals in the upper deciles.

5. *7th to bottom decile stars:* These stars hardly move within the circle across the lower deciles (6th–bottom). This indicates the individuals across the least responsive deciles are essentially the same.

In sum, the ten many-variables star graphs provide a lifelike animation of how the four-dimensional profiles of the individuals change as the model assigns the individuals into the ten deciles. The first five deciles show a slight progressive decrease in education and age means. Between the fifth and sixth deciles, there is a sudden change in full profile, as the gender contribution to profile is now skewed to males. Across the bottom five deciles, there is a slight progressive decrease in income and age means.

14.5 Profile Curves Method

I present the Profile Curves Method as an alternative geometry-based graphical method for the problem previously tackled by the star graphs, but from a slightly different concern. The star graphs provide the marketer with strategic marketing intelligence for campaign development, as obtained from an ocular inspection and complete descriptive account of their customers in terms of variables of interest. In contrast, the profile curves provide the marketer with strategic marketing intelligence for model implementation, specifically, determining the number of reliable decile groups.

The profile curves are demanding to conceptually build and interpret, unlike the star graphs. Its construction is not intuitive, as it uses a series of unanticipated trigonometric functions, and its display is disturbingly abstract. However, the value of profile curves in a well-matched problem-solution (as presented here) can offset the initial reaction and difficulty in its use.

I discuss below the basics of profile curves, and the *profile analysis*, which serves as a useful preliminary step to the implementation of the Profile Curves Method. I illustrate the profile analysis-profile curves methodical order using the RESPONSE model decile analysis table. The profile analysis involves simple pairwise scatterplots. The Profile Curves Method requires a special computer program, which SAS/Graph has. I provide the SAS code for the profile curves at the end of the chapter for the illustration presented.

14.5.1 Profile Curves[3] Basics

Consider the curve function $f(t)$ defined in Equation (14.1): [9]

$$f(t) = X_1 / \sqrt{2} + X_2 sin(t) + X_3 cos(t) + X_4 sin(2t) + X_5 cos(2t) + ... \qquad (14.1)$$

where $$-\pi \leq t \leq \pi$$

The curve function $f(t)$ is a *weighted* sum of *basic curves* for an observation X, which is represented by many variables, i.e., a multivariate data array, X = {$X_1, X_2, X_3, ..., X_k$}. The *weights* are the values of the X's. The *basic curves* are trigonometric functions sine and cosine. The plot of $f(t)$ on the y-axis and t on the x-axis, for a set of multivariate data arrays (rows) of mean values for *groups* of individuals are called *profile curves*.

Like the star graphs, the profile curves are a visual display of multivariate data, especially effective for a small table, like the decile analysis table. Unlike the star graphs, which provide visual displays for single and many variables jointly, the profile curves only provide a visual display of the *joint effects* of the X variables across several groups. A profile curve for a single group is an abstract mathematical representation of the row of mean values of the variables. As such, a single group curve imparts no usable information.

---

[3] Profile curves are also known as curve plots.

Information is extracted from a comparative evaluation of two or more group curves. Profile curves permit a qualitative assessment of the differences among the persons across the groups. In other words, profile curves serve as a method of multivariate comparison of groups.

14.5.2 Profile Analysis

Database marketers use models to classify customers into ten deciles, ranging from the top 10% most-likely to perform to the bottom 10% least-likely to perform. To communicate effectively to the customers, database marketers combine the deciles into groups, typically three, top, middle and bottom *decile groups*. The top group typifies high-valued/high-responsive customers to "harvest" by sending provoking solicitations for preserving their performance levels. The middle group represents medium-valued/medium-responsive customers to "retain and grow" by including them in marketing programs tailored to keep and further stimulate their performance levels. Last, the bottom group depicts a segment of minimal performers and former customers, whose performance levels can be rekindled and reactivated by creative new product and discounted offerings.

Profile analysis is used to create decile groups. Profile analysis consists of a) calculating the means of the variables of interest, and b) plotting the means of several *pairs* of the variables. These *profile plots* suggest how the individual deciles may be combined. However, because the profiles are multidimensional (i.e., defined by many variables), the assessment of the many profile plots provides an incomplete view of the groups. If the profile analysis is fruitful, it serves as guidance for the profile curves method in determining the number of reliable decile groups.

14.6 Illustration

Returning to the RESPONSE model decile analysis table (Table 14.1), I construct three profile plots in Figures 14.3 to 14.5: AGE with both INCOME and GENDER, and EDUCATION with INCOME. The AGE–INCOME plot indicates that AGE and INCOME jointly decrease down through the deciles. That is, the most responsive customers are older with more income, and the least responsive customers are younger with less income.

The AGE–GENDER plot shows that the most responsive customers are older and prominently female, and the least responsive customers are younger and typically male. The EDUCATION–INCOME plot shows that the most responsive customers are better educated with higher incomes.

Similar plots for the other pairs of variables can be constructed, but the task of interpreting all the pairwise plots is formidable.[4]

---

[4] Plotting three variables at a time can be done by plotting a pair of variables for each decile value of the third variable: clearly, a challenging effort.

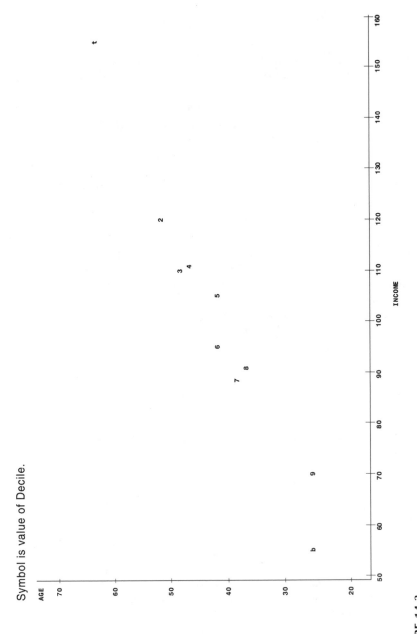

Symbol is value of Decile.

FIGURE 14.3
Plot of AGE and INCOME. t = top; b = bottom.

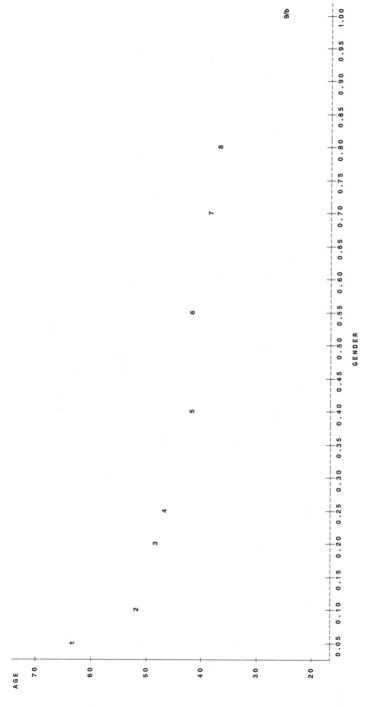

FIGURE 14.4
Plot of AGE and GENDER by Decile. t = top; b = bottom.

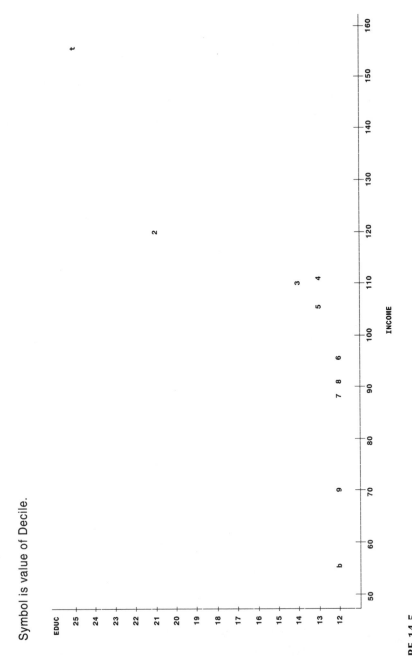

FIGURE 14.5

Plot of EDUCATION and INCOME by Decile. t = top; b = bottom.

The three profile plots indicate various candidate decile-group compositions:

1. The AGE–INCOME plot suggests defining the groups as follows:
    i. top group — top decile;
    ii. middle group — 2nd through 8th deciles;
    iii. bottom group — 9th and bottom deciles.
2. The AGE–GENDER plot does not reveal any grouping.
3. The EDUCATION–INCOME plot indicates:
    i. top group — top decile;
    ii. middle group — 2nd decile;
    iii. bottom group — 3rd through bottom deciles. Actually, the latter group can be divided into two subgroups: deciles 3 to 8, and deciles 9 and bottom.

In this case, the profile analysis is not fruitful. It is not clear how to define the best decile grouping based on the above findings. Additional plots would probably produce other inconsistent findings.

14.6.1 Profile Curves for RESPONSE Model

The Profile Curves Method provides a graphical presentation, in Figure 14.6, of the joint effects of the four demographic variables across all the deciles based on the decile analysis in Table 14.1. Interpretation of this graph is part of the following discussion, in which I illustrate the strategy for creating reliable decile groups with profile curves.

Under the working assumption that the top/middle/bottom decile groups exist, I create profile curves for the top, 5th and bottom deciles, in Figure 14.7, as defined below in Equations (14.2), (14.3) and (14.4), respectively.

$$f(t)_{top\_decile} = 63 / \sqrt{2} + 155sin(t) + 18cos(t) + .05sin(2t) \tag{14.2}$$

$$f(t)_{5th\_decile} = 42 / \sqrt{2} + 105sin(t) + 13cos(t) + .40sin(2t) \tag{14.3}$$

$$f(t)_{bottom\_decile} = 25 / \sqrt{2} + 55sin(t) + 12cos(t) + 1.00sin(2t) \tag{14.4}$$

The upper, middle and lower profile curves correspond to the rows of means for the top, 5th and bottom deciles, respectively, which are reproduced for ease of discussion in Table 14.2. The three profile curves form two "hills." Based on a subjective assessment,[5] I declare that the profile curves are different in terms of slope of the hill. The implication is that the individuals in

---

[5] I can test for statistical difference (see Andrew's article); however, I am doing a visual assessment, which does not require any statistical rigor.

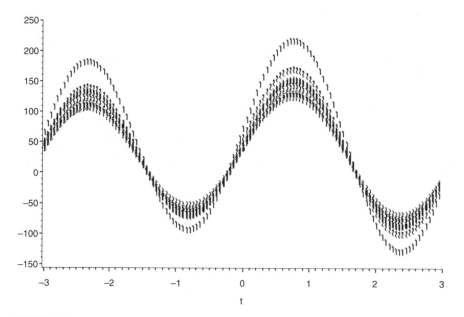

FIGURE 14.6
Profile Curves: All Deciles

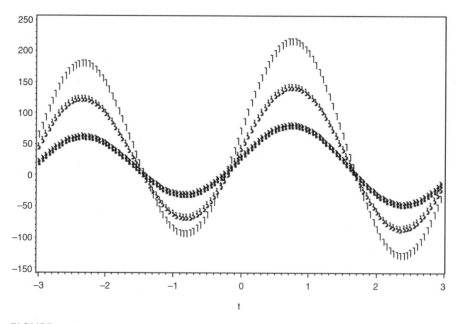

FIGURE 14.7
Profile Curves for Top, 5th and Bottom Deciles

TABLE 14.2

Response Decile Analysis: Demographic Means for Top, 5, and Bottom Deciles

Decile	AGE (yrs)	INCOME ($000)	EDUCATION (Years of Schooling)	GENDER (1 = Male/ 0 = Female)
Top	63	155	18	0.05
5	42	105	13	0.40
Bottom	25	55	12	1.00

each decile are different — with respect to the four demographic variables considered jointly — from the individuals in each of the other two deciles, and consequently the deciles cannot be combined.

The three-profile curves graph in Figure 14.7 exemplifies how the profile curves method works. Large variation among rows corresponds to a set of profile curves that greatly depart from a common shape. Disparate profile curves indicate that the rows should remain separate, not combined. Profile curves that slightly depart from a common shape indicate that the rows can be combined to form a more reliable row. Accordingly, I restate my preemptive implication: the three-profile curves graph indicates that individuals across the three deciles are diverse with respect to the four demographic variables considered jointly, and the deciles cannot be aggregated to form a homogeneous group.

When the rows are obviously different, as in the case in Table 14.2, the profile curves method serves as a confirmatory method. When the variation in the rows are not apparent, as is likely with a large number of variables, noticeable row-variation is harder to discern, in which case the profile curves method serves as an exploratory tool.

14.6.2 Decile-Group Profile Curves

I have an initial set of decile groups: the top group is the top decile; the middle group is the 5th decile; and the bottom group is the bottom decile. I have to assign the remaining deciles to one of the three groups. Can I include the 2nd decile in the top group? The answer lies in the graph for the top and 2nd deciles in Figure 14.8, from which I observe that the top and 2nd decile profile curves are different. Thus, the top group remains with only the top decile. The 2nd decile is assigned to the middle group; more about this assignment later.

Can I add the 9th decile to the bottom group? From the graph for the 9th and bottom deciles in Figure 14.9, I observe that the two profile curves are not different. Thus, I add the 9th decile to bottom group. Can the 8th decile also be added to the bottom group (now consisting of the 9th and bottom deciles)? I observe in the graph for the 8th–bottom deciles in Figure 14.10 that the 8th decile profile curve is somewhat different from the 9th and

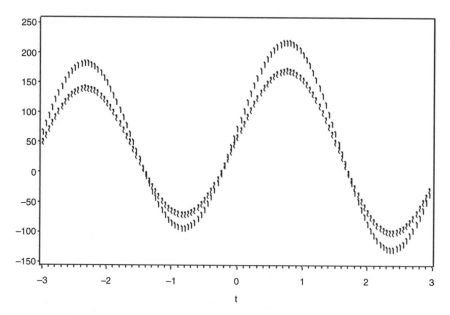

FIGURE 14.8
Profile Curves for Top, and 2nd Deciles

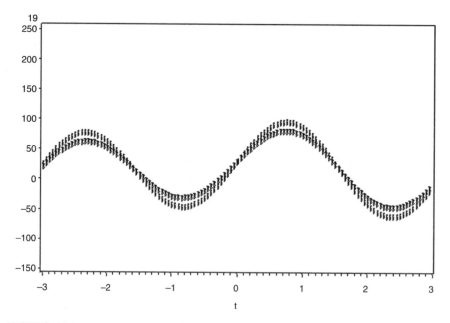

FIGURE 14.9
Profile Curves for 9th to Bottom Deciles

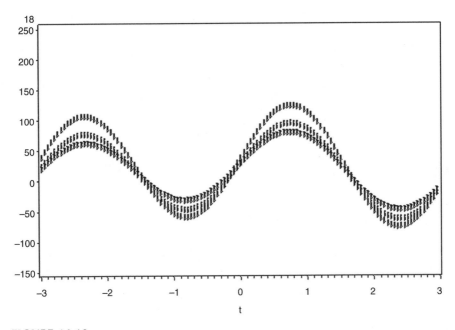

FIGURE 14.10
Profile Curves for 8th to Bottom Deciles

bottom decile profile curves. Thus, I do not include the 8th decile to the bottom group; it is placed in the middle group.

To insure that the middle group can be defined by combining deciles 2 through 8, I generate the corresponding graph in Figure 14.11. I observe a bold common curve formed by the seven profile curves tightly stacked together. This suggests that the individuals across the deciles are similar. I conclude that the group comprised of deciles 2 through 8 is homogeneous.

Two points are worth noting: first, the density of the middle group's common curve is a measure of homogeneity among the individuals across these middle group deciles. Except for a little "daylight" at the top of the right-side hill, the curve is solid. This indicates that nearly all the individuals within the deciles are alike. Second, if the curve had a *pattern* of daylight then the middle group would be divided into subgroups defined by the pattern.

In sum, I define the final decile groups as follows:

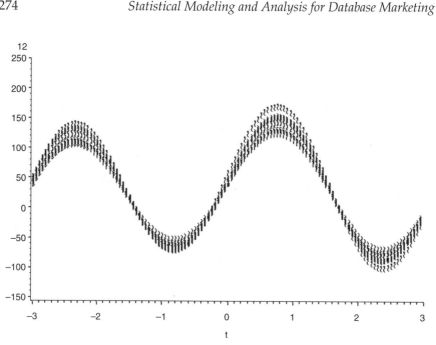

FIGURE 14.11
Profile Curves 2nd through 8th Deciles

1. Top group — top decile
2. Middle group — 2nd through 8th deciles
3. Bottom group — 9th and bottom deciles

The sequential use of profile curves (as demonstrated here) serves as a general technique for multivariate profile analysis. In some situations where the profile data appear obvious, a single graph with all ten decile curves may suffice. For the present illustration, the single graph in Figure 14.6 shows patterns of daylight that suggest the final decile groups. A closer look at the two hills confirms the final decile groups.

14.7 Summary

After taking a quick look back at the history of the graph, from its debut as an Egyptian map to today's visual displays in three-dimensional with 48-bit color, I focused on two underused methods of displaying multivariate data. I proposed the star graphs and profile curves as methods of visualizing "what the final model is doing" upon implementation of its intended task of predicting.

I presented the basics of star graph construction along with an illustration of a response model with four demographic variables. The utility of the star graph was made apparent when I compared its results with those of the traditional profiling, such as examining the means of variables of interest across the deciles. Traditional profiling provides good information, but it is neither compelling nor complete in its examination of how the response model works. It is incomplete in two respects: 1) it only considers one variable at a time without any display; and 2) it does not beneficially stimulate the marketer into strategic thinking for insights. The star graph with its unique visual display does stimulate strategic thinking. However, like the traditional profiling, the star graph only considers one variable at a time.

Accordingly, I extended the single-variable star graph application to the many-variable star graph application. Continuing with the response illustration, I generated many-variable star graphs, which clearly provided the "beneficial stimulation" desired. The many-variable star graphs displayed a full profile of the individuals using all four demographic variables considered jointly for a complete understanding of how the response model works as it assigns individuals into the response deciles.

Last, I presented the Profile Curves Method as an alternative method for the problem previously tackled by the star graphs, but from a slightly different concern. The star graph provides the marketer with strategic intelligence for campaign development. In contrast, the profile curves provide the marketer with strategic information for model implementation, specifically, determining the number of reliable decile groups.

The first-ever encounter with profile curves can be unremittingly severe on the data analyst: they are demanding, not only in their construction but in their interpretation. I demonstrated with the response model illustration that the long term utility of profile curves offsets its initial affront to the analytical senses. The response model profile curves, which accompanied a detailed analysis and interpretation, clearly showed the determination of the reliable decile groups.

14.8 SAS Code for Star Graphs for Each Demographic Variable about the Deciles

```
Title1 'table';
data table;
input decile age income educ gender;
cards;
1    63 155 18 0.05
2    51 120 16 0.10
```

```
3     49 110 14 0.20
4     46 111 13 0.25
5     42 105 13 0.40
6     41 095 12 0.55
7     39 088 12 0.70
8     37 091 12 0.80
9     25 070 12 1.00
10    25 055 12 1.00
;
run;
proc print;
run;
proc standard data = table out = tablez
mean = 4 std = 1;
var
age income educ gender;
Title1 'table stdz';
proc print data = tablez;
run;
proc format; value dec_fmt
1. = 'top' 2 = ' 2 ' 3 = ' 3 ' 4 = ' 4 ' 5 = ' 5 '
6. = ' 6 ' 7 = ' 7 ' 8 = ' 8 ' 9 = ' 9 ' 10 = 'bot';
run;
proc greplay nofs igout = work.gseg;
delete all;
run;quit;
goptions reset = all htext = 1.05 device = win
targetdevice = winprtg ftext = swissb lfactor = 3
hsize = 2 vsize = 8;
proc greplay nofs igout = work.gseg;
delete all;
run;
goptions reset = all device = win
targetdevice = winprtg ftext = swissb lfactor = 3;
title1 'AGE by Decile';
proc gchart data = tablez;
format decile dec_fmt. ;
```

```
star decile/fill = empty discrete sumvar = age
slice = outside value = none noheading ;
run;quit;
Title1 'EDUCATON by Decile';
proc gchart data = tablez;
format decile dec_fmt. ;
star decile/fill = empty discrete sumvar = educ
slice = outside value = none noheading;
run;quit;
Title1 'INCOME by Decile';
proc gchart data = tablez;
format decile dec_fmt. ;
star decile/fill = empty discrete sumvar = income
slice = outside value = none noheading;
run;quit;
Title1 'GENDER by Decile';
proc gchart data = tablez;
format decile dec_fmt.;
star decile/fill = empty discrete sumvar = gender
slice = outside value = none noheading;
run;quit;
proc greplay nofs igout = work.gseg tc = sashelp.templt
template = 12r2s;
treplay 1:1 2:2 3:3 4:4;
run;quit;
```

14.9 SAS Code for Star Graphs for Each Decile about the Demographic Variables

```
data table;
input decile age income educ gender;
cards;
1    63 155 18 0.05
2    51 120 16 0.10
3    49 110 14 0.20
```

```
4    46 111 13 0.25
5    42 105 13 0.40
6    41 095 12 0.55
7    39 088 12 0.70
8    37 091 12 0.80
9    25 070 12 1.00
10   25 055 12 1.00
;
run;
proc standard data = table out = tablez
mean = 4 std = 1;
var
age income educ gender;
Title2 'table stdz';
proc print data = tablez;
run;
proc transpose data = tablez out = tablezt prefix =
dec_;
var
age income educ gender;
run;
proc print data = tablezt;
run;
proc standard data = tablezt out = tableztz
mean = 4 std = 1;
Var
dec_1 - dec_10;
title2'tablezt stdz';
proc print data = tableztz;
run;
proc transpose data = tablez out = tablezt prefix =
dec_;
var
age income educ gender;
run;
proc print data = tablezt;
run;
```

```
proc greplay nofs igout = work.gseg;
delete all;
run;quit;
goptions reset = all htext = 1.05 device = win
target = winprtg ftext = swissb lfactor = 3
hsize = 4 vsize = 8;
Title1 'top decile';
proc gchart data = tableztz;
star name/fill = empty sumvar = dec_1
slice = outside value = none noheading;
run;quit;
Title1 '2nd decile';
proc gchart data = tableztz;
star name/fill = empty sumvar = dec_2
slice = outside value = none noheading;
run;quit;
Title1 '3rd decile';
proc gchart data = tableztz;
star name/fill = empty sumvar = dec_3
slice = outside value = none noheading;
run;quit;
Title1 '4th decile';
proc gchart data = tableztz;
star name/fill = empty sumvar = dec_4
slice = outside value = none noheading;
run;quit;
Title1 '5th decile';
proc gchart data = tableztz;
star name/fill = empty sumvar = dec_5
slice = outside value = none noheading;
run;quit;
Title1 '6th decile';
proc gchart data = tableztz;
star name/fill = empty sumvar = dec_6
slice = outside value = none noheading;
run;quit;
Title1 '7th decile';
```

```
proc gchart data = tableztz;
star name/fill = empty sumvar = dec_7
slice = outside value = none noheading;
run;quit;
Title1 '8th decile';
proc gchart data = tableztz;
star name/fill = empty sumvar = dec_8
slice = outside value = none noheading;
run;quit;
Title1 '9th decile';
proc gchart data = tableztz;
star name/fill = empty sumvar = dec_9
slice = outside value = none noheading;
run;quit;
Title1 'bottom decile';
proc gchart data = tableztz;
star name/fill = empty sumvar = dec_10
slice = outside value = none noheading;
run;quit;
goptions hsize = 0 vsize = 0;
Proc Greplay Nofs TC = Sasuser.Templt;
Tdef L2R5 Des = 'Ten graphs: five across, two down'
1/llx = 0 lly = 51
ulx = 0 uly = 100
urx = 19 ury = 100
lrx = 19 lry = 51
2/llx = 20 lly = 51
ulx = 20 uly = 100
urx = 39 ury = 100
lrx = 39 lry = 51
3/llx = 40 lly = 51
ulx = 40 uly = 100
urx = 59 ury = 100
lrx = 59 lry = 51
4/llx = 60 lly = 51
ulx = 60 uly = 100
urx = 79 ury = 100
```

```
lrx = 79 lry = 51
5/llx = 80 lly = 51
ulx = 80 uly = 100
urx = 100 ury = 100
lrx = 100 lry = 51
6/llx = 0 lly = 0
ulx = 0 uly = 50
urx = 19 ury = 50
lrx = 19 lry = 0
7/llx = 20 lly = 0
ulx = 20 uly = 50
urx = 39 ury = 50
lrx = 39 lry = 0
8/llx = 40 lly = 0
ulx = 40 uly = 50
urx = 59 ury = 50
lrx = 59 lry = 0
9/llx = 60 lly = 0
ulx = 60 uly = 50
urx = 79 ury = 50
lrx = 79 lry = 0
10/llx = 80 lly = 0
ulx = 80 uly = 50
urx = 100 ury = 50
lrx = 100 lry = 0;
Run;
Quit;
Proc Greplay Nofs Igout = Work.Gseg
TC = Sasuser.Templt Template = L2R5;
Treplay 1:1 2:2 3:3 4:4 5:5 6:6 7:7 8:8 9:9 10:10;
run;quit;
```

14.10 SAS Code for Profile Curves: All Deciles

```
Title1'table';
```

```
data table;
input decile age income educ gender;
cards;
1 63 155 18 0.05
2 51 120 16 0.10
3 49 110 14 0.20
4 46 111 13 0.25
5 42 105 13 0.40
6 41 095 12 0.55
7 39 088 12 0.70
8 37 091 12 0.80
9 25 070 12 1.00
10 25 055 12 1.00
;
run;
data table;
Set table;
x1 = age;x2 = income;x3 = educ;x4 = gender;
proc print;
run;
data table10;
sqrt2 = sqrt(2);
array f {10};
do t = -3.14 to 3.14 by.05;
do i = 1 to 10;
set table point = i;
f(i) = x1/sqrt2 + x4*sin(t) + x3*cos(t) + x2*sin(2*t);
end;
output;
label f1 = '00'x;
end;
stop;
run;
goptions reset = all device = win target = winprtg ftext
= swissb lfactor = 3;
Title1 'Figure 14.6 Profile Curves: All Deciles';
proc gplot data = table10; plot
```

```
f1*t = 'T'
f2*t = '2'
f3*t = '3'
f4*t = '4'
f5*t = '5'
f6*t = '6'
f7*t = '7'
f8*t = '8'
f9*t = '9'
f10*t = 'B'
/overlay haxis = -3 -2 -1 0 1 2 3
nolegend vaxis = -150 to 250 by 50;
run;quit;
```

References

1. Descartes, R., *The Geometry of Rene Descartes*, Dover, New York, 1954.
2. Costigan-Eaves, P., *Data Graphics in the 20th Century: A Comparative and Analytical Survey.* Ph.D. thesis, Rutgers University, New Jersey, 1984.
3. Funkhouser, H.G., Historical development of the graphical representation of statistical data, *Osiris*, 3, 269–404, 1937.
4. Du Toit, S.H.C., Steyn, A.G.W., and Stumpf, R.H., *Graphical Exploratory Data Analysis*, Springer-Verlag, New York, 1986, p.2.
5. Tukey, J.W., *The Exploratory Data Analysis*, Addison-Wesley, Reading, MA, 1977.
6. Salsburg, D., *The Lady Tasting Tea*, Freeman, New York, 2001.
7. Snee, R.D., Hare, L.B., and Trout, J. R., *Experiments in Industry. Design, Analysis and Interpretation of Results*, American Society for Quality, Milwaukee, WI, 1985.
8. Friedman, H.P., Farrell, E.S., Goldwyn, R.M., Miller, M., and Siegel, J., A graphic way of describing changing multivariate patterns, *Proceedings of the Sixth Interface Symposium on Computer Science and Statistics*, University of California Press, Berkeley, 1972.
9. Andrews, D.F., Plots of high-dimensional data, *Biometrics*, 28, 1972.

15

Genetic Modeling in Database Marketing: The GenIQ Model[1]

Using a variety of techniques, data analysts in database marketing aim to build models that maximize expected response and profit from solicitations. Standard techniques include the statistical methods of classical discriminant analysis, as well as logistic and ordinary regression. A recent addition in the data analysis arsenal is the machine learning (ML) method of neural networks. The GenIQ Model is a hybrid ML-statistics method that is presented in full detail below.

First, a background on the concept of optimization will be helpful since optimization techniques provide the estimation of all models. Genetic modeling is the "engine" for the GenIQ Model, and is discussed next as an ML optimization approach. Since the objectives of database marketing are to maximize expected response and profit from solicitations, I will demonstrate how the GenIQ Model serves to meet those objectives. Actual case studies will further explicate the potential of the GenIQ Model.

15.1 What Is Optimization?

Whether in business or in model building, optimization is central to the decision-making process. In both theory and method, optimization involves selecting the best, or most favorable condition within a given environment. To distinguish among available choices, an objective (or fitness function) must be determined. Whichever choice corresponds to the extreme value[2] of the objective function is the best alternative, and thus the solution to the problem.

---

[1] This chapter is based on an article in *Journal of Targeting, Measurement and Analysis for Marketing*, 9, 3, 2001. Used with permission.
[2] If the optimization problem seeks to minimize the objective function, then the extreme value is the smallest; if it seeks to maximize, then the extreme value is the largest.

Modeling techniques are developed to find a specific solution to a problem. In database marketing, one such problem is to predict sales. The least-squares regression technique is a model formulated to address sales prediction. The regression problem is formulated in terms of finding the regression equation such that the prediction errors (the difference between actual and predicted sales) are small.[3] The objective function is the prediction error, making the best equation the one that minimizes that prediction error. Calculus-based methods are used to estimate the best regression equation.

As I will later discuss, each modeling method addresses its own decision problem. The GenIQ model addresses a problem specific to database marketing, and uses genetic modeling as the optimization technique for its solution.

15.2 What Is Genetic Modeling?

Just as Darwin's principle of the survival of the fittest explains tendencies in human biology, database marketers can use the same principle to predict the best current solution to an optimization problem.[4] Each genetic model has an associated fitness value that indicates how well the model solves, or "fits" the problem. A model with a high fitness value solves the problem better than a model with a lower fitness value, and survives and reproduces at a high rate. Models that are less fit survive and reproduce, if at all, at a lower rate.

If two models are effective in solving a problem, then some of their parts undoubtedly contain some valuable genetic material. Recombining the parts of highly fit parent models can sometimes produce offspring models that are better fit at solving the problem than either parent. Offspring models then become the parents of the next generation, repeating the recombination process. After many generations, an evolved model is declared the best-so-far solution of the problem.

Genetic modeling[5] consists of the following steps: [1]

1. Definition of the fitness function. The fitness function allows for identifying good or bad models, after which refinements are made with the goal of producing the best model.

---

[3] The definition of "small" (technically called mean squared error) is the average of the squared differences between actual and predicted values.

[4] The focus of this chapter is optimization, but genetic modeling has been applied to a variety of problems: optimal control, planning, sequence induction, empirical discovery and forecasting, symbolic integration and discovering mathematical identities.

[5] Genetic modeling as described in this chapter is formally known as genetic programming. I choose the term "modeling" instead of "programming" because the latter term, which has its roots in computer sciences, does not connote the activity of model building to data analysts with statistics or quantitative backgrounds.

2. Selection of a set of functions (e.g., the set of arithmetic operators {addition, subtraction, multiplication, division}; log and exponential) and variables (predictors X_1, X_2, ..., X_n, and numerical constants) believed to be related to the problem at hand (the dependent variable Y).[6] An initial population of random models must be generated using the preselected set of functions and variables.

3. Calculation of the fitness of each model in the population by applying the model to a training set, a sample of individuals along with their values on the predictor variables, X_1, X_2, ..., X_n and the dependent variable Y. Thus, every model has a fitness value reflecting how well it solves the problem.

4. Creation of a new population of models by applying the following three operations. The operations are applied to models in the current population selected with a probability based on fitness (i.e., the fitter the model, the more likely the model is to be selected).

   a. Reproduction: Copying an existing model to the new population.

   b. Crossover: Creation of two offspring models for the new population by genetically recombining randomly chosen parts of two existing parent models.

   c. Mutation: Introduction of random changes in some models.

The model with the highest fitness value produced in a generation is declared the best-of-generation model, which may be the solution, or an approximate solution to the problem.

15.3 Genetic Modeling: An Illustration

Let's look at the proccess of building a response model. I will designate the best model as one with the highest R-squared[7] value. Thus, the fitness function is the formula for R-squared. (Analytical note: I am using the R-squared measure for only illustration purposes.)

The modeler has to decide on functions and variables that are related to the problem at hand (e.g., predicting response). Unless there is theoretical justification, empirical experience — usually based on trial-and-error — provides the guidance for the functions and variables.

I have two variables, X_1 and X_2, to use as predictors of response. Thus, the variable set contains X_1 and X_2. I add the numerical constant "b" to the

---

[6] Effectively, I have chosen a genetic alphabet.

[7] I know that R-squared is not an appropriate fitness function for a 0-1 dependent variable model. Perhaps I should use the likelihood function of the logistic regression model as the fitness measure for the response model, or an example with a continuous (profit) variable and the R-squared fitness measure. More about the appropriate choice of fitness function for the problem at hand in a later section.

variable set based on prior experience. I define the function set to contain the four arithmetic operations and the exponential function (exp), also based on prior experience.

Generating the initial population is done with an *unbiased function roulette wheel*, in Figure 15.1, and an *unbiased function-variable roulette wheel*, in Figure 15.2. The slices of the function wheel are of equal size, namely, 20%. The slices of the function-variable wheel are of equal size, namely, 12.5%. Note that the division symbol "%" is used to denote a "protected" division. This means that division by zero is set to the value 1.

To generate the first random model, I spin the function wheel. The wheel's pointer lands on slice "+." Next, I spin the function-variable wheel; the pointer lands on slice X1. From the next two spins of the function-variable

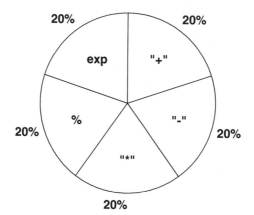

FIGURE 15.1
Unbiased Function Roulette Wheel

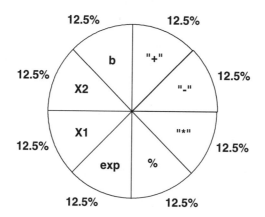

FIGURE 15.2
Unbiased Function-Variable Roulette Wheel

wheel, the pointer lands on slices "+" and "b," successively. I decide to stop evolving the model at this point. The resultant random model 1 is depicted, in Figure 15.3, as a *rooted point-label tree*.

I generate the second random model, in Figure 15.4, by spinning the function wheel once, then by spinning the function-variable wheel twice. The pointer lands on slices "+," X_1 and X_1, successively. Similarly, I generate three more random models, models 3, 4 and 5 in Figures 15.5, 15.6 and 15.7. Thus, I have generated the initial population of five random models (population size is five).

Each of the five models in the population is assigned a fitness value, in Table 15.1, indicating how well it solves the problem of predicting response. Because I am using R-squared as the fitness function, I apply each model to a training dataset and calculate its R-squared value. Model 1 produces the

```
Response
   |
   +
  / \
 b   X1
```

Model 1: Response = b + X1

FIGURE 15.3
Random Model 1

```
Response
   |
   +
  / \
X1   X1
```

Model 2: Response = X1 + X1

FIGURE 15.4
Random Model 2

```
Response
   |
   *
  / \
X1   X1
```

Model 3: Response = X1 * X1

FIGURE 15.5
Random Model 3

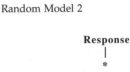

```
Response
   |
   *
  / \
X1   +
    / \
   b   X2
```

Model 4: Response = X1 * (b + X2)

FIGURE 15.6
Random Model 4

```
Response
   |
   *
  / \
X1   exp
      |
      X2
```

Model 5: Response = X1 *exp(X2)

FIGURE 15.7
Random Model 5

TABLE 15.1

Initial Population

	Fitness Value (R-Squared)	PTF (Fitness/Total)
Model 1	0.52	0.34
Model 2	0.41	0.28
Model 3	0.38	0.25
Model 4	0.17	0.11
Model 5	0.05	0.03
Population total fitness	1.53	

highest R-squared value, 0.52, and model 5 produces the lowest R-squared value, 0.05.

Fitness for the population itself can be calculated. The *total fitness of the population* is the sum of the fitness values among all models in the population. Here, the population total fitness is 1.53, in Table 15.1.

15.3.1 Reproduction

After the initial population is generated, the operation of reproduction takes place. Reproduction is the process by which models are duplicated or copied based on *selection proportional to fitness* (PTF). PTF is defined as: model fitness value divided by population total fitness (see Table 15.1.) For example, model 1 has a PTF value of 0.34 (= 0.52/1.53).

Reproduction PTF means that a model with a high PTF value has a high probability of being selected for inclusion in the next generation. The reproduction operator is implemented with a *biased model roulette wheel*, in Figure 15.8, where the slices are sized according to PTF values.

The operation of reproduction proceeds as follows. The spin of the biased model roulette wheel determines which and how many models are copied. The model selected by the pointer is copied without alteration and put into the next generation. Spinning the wheel in Figure 15.8, say 100 times, produces on average the selection: 34 copies of model 1, 28 copies of model 2, 25 copies of model 3, 11 copies of model 4, and 3 copies of model 5.

15.3.2 Crossover

The crossover (sexual recombination) operation is performed on two parent models by genetically recombining randomly chosen parts of the two existing parent models, the expectation being that the offspring models are even more fit than either parent model.

The crossover operation works with selection PTF. An illustration makes this operation easy to understand. Consider the two parent models in Figures

15.9 and 15.10. The operation begins by randomly selecting an internal point (a function) in the tree for the *crossover site.*

Say, the crossover sites are the lower "+" and "*" for parents 1 and 2, respectively.

The *crossover fragment* for a parent is the subtree that has at its root the crossover site function. Crossover fragments for parents 1 and 2 are in Figures 15.11 and 15.12, respectively.

Offspring 1, in Figure 15.13, from parent 1 is produced by deleting the crossover fragment of parent 1 and then inserting the crossover fragment of parent 2 at the crossover point of the parent 1. Similarly, offspring 2, in Figure 15.14, is produced.

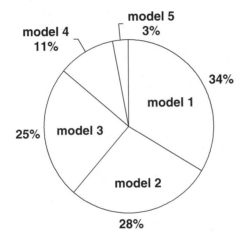

FIGURE 15.8
Biased Model Roulette Wheel

Parent 1

```
   +
  / \
X2   +
    / \
   b   X1
```

FIGURE 15.9
Parent 1

Parent 2

```
      -
     / \
    *   X4
   / \
  c   X3
```

FIGURE 15.10
Parent 2

Crossover Fragment 1

```
  +
 / \
b   X1
```

FIGURE 15.11
Crossover Fragment 1

Crossover Fragment 2

```
  *
 / \
c   X3
```

FIGURE 15.12
Crossover Fragment 2

Offspring 1
```
        +
      /   \
    X2     *
          / \
         c   X3
```

FIGURE 15.13
Offspring 1

Offspring 2
```
        -
      /   \
    +      X4
   / \
  b   X1
```

FIGURE 15.14
Offspring 2

15.3.3 Mutation

The mutation operation begins by selecting a point at random within a tree. This mutation point can be an internal point (a function) or an external or terminal point (a variable or numerical constant). The mutation operation either *replaces* a randomly generated function with another function (from the function set previously defined), or *inverts*[8] the terminals of the subtree whose root is the randomly selected internal point.

For example, model I, in Figure 15.15, is mutated by replacing the function "−" with "+," resulting in mutated model I.1 in Figure 15.16. Model I is also mutated by inverting the terminal points, c and X3, resulting in mutated model I.2 in Figure 15.17.

Model I
```
        +
      /   \
    X1     -
          / \
         c   X3
```

FIGURE 15.15
Model I for imputation

Model I.1
```
        +
      /   \
    X1     +
          / \
         c   X3
```

FIGURE 15.16
Mutated Model I.1

Model I.2
```
        +
      /   \
    X1     -
          / \
         X3   c
```

FIGURE 15.17
Muted Model I.2

---

[8] When a subtree has more than two terminals, the terminals are randomly permutated.

15.4 Parameters for Controlling a Genetic Model Run

There are several control parameters that need to be set before evolving a genetic model.

1. Population size, or the number of models randomly generated and subsequently evolved.
2. The maximum number of generations to be run.
3. Reproduction probability, or the percentage of the population which is copied. If population size is 100 and reproduction probability is 10%, then 10 models from each generation are selected (with reselection allowed) for reproduction. Selection is based on PTF.
4. Crossover probability, or the percentage of the population which is used for crossover. If population size is 100 and crossover probability is 80%, then 80 models from each generation are selected (with reselection allowed) for crossover. Selection is based on PTF. Models are paired at random.
5. Mutation probability, or the percentage of the population which is used for mutation. If population size is 100 and mutation 10%, then 10 models from each generation are selected (with reselection allowed) for mutation. Selection is based on PTF.
6. Termination Criterion, or the single model with the largest fitness value over all generations, the so-called best-so-far model, is declared the result of a run.

15.5 Genetic Modeling: Strengths and Limitations

Genetic Modeling has strengths and limitations, like any methodology. Perhaps, the most important strength of genetic modeling is that it is a workable alternative to traditional methods, which are highly parametric with sample size restrictions. Traditional methods have algorithms depending on smooth, unconstrained functions with the existence of derivatives (well-defined slope values). In practice, the functions (response surfaces) are noisy, multimodal and frequently discontinuous. In contrast, genetic models are robust, assumption-free, nonparametric models, and perform well on large and small samples. The only requirement is a fitness function, which can be designed to insure that the genetic model does not perform worse than any other method.

Genetic Modeling has shown itself to be effective for solving large optimization problems and searching very large datasets. In addition, genetic modeling can be used to learn complex relationships, making it a viable data mining tool for rooting out valuable pieces of information.

A potential limitation of genetic modeling is in the setting of the genetic modeling parameters: population size, and reproduction, crossover and mutation probabilities. The parameter settings are, in part, data and problem dependent, thus proper settings require experimentation. Fortunately, new theories and empirical studies are continually providing rules-of-thumb for these settings as application areas broaden. These guidelines make genetic modeling an accessible approach to analysts not formally trained in genetic modeling. Even with the "correct" parameter settings, genetic models do not guarantee the optimal (best) solution. Further, genetic models are only as good as the definition of the fitness function. Precisely defining the fitness function sometimes requires expert experimentation.

15.6 Goals of Modeling in Database Marketing

Database marketers typically attempt to improve the effectiveness of their solicitations by targeting their best customers or prospects. They use a model to identify individuals who are likely to respond to or generate profit[9] from a solicitation. The model provides, for each individual, estimates of probability of response and estimates of contribution-to-profit. Although the precision of these estimates is important, the model's performance is measured at an aggregated level as reported in a decile analysis.

Database marketers have defined the *Cum Lift*, which is found in the decile analysis, as the relevant measure of model performance. Based on the model's selection of individuals, database marketers create a solicitation list to obtain an advantage over a random selection of individuals. The Cum Response Lift is an index of how many more responses are expected with a selection based on a model over the expected responses with a random selection (without a model). Similarly, the Cum Profit Lift is an index of how much more profit is expected with a selection based on a model over the profit expected with a random selection (without a model). The concept of Cum Lift and the steps of the construction in a decile analysis are described in Chapter 12.

It should be clear that a model that produces a decile analysis with more responses or profit in the upper (top, 2, 3 or 4) deciles is a better model than a model with less responses or profit in the upper deciles. This concept is the motivation for the GenIQ Model.

---

[9] I use the term "profit" as a stand-in for any measure of an individual's worth, such as sales per order, lifetime sales, revenue, number of visits or number of purchases.

15.7 The GenIQ Response Model

The GenIQ approach to modeling is to specifically address the objectives concerning database marketers, namely, maximizing response and profit from solicitations. The GenIQ Model uses the genetic methodology to *explicitly* optimize the desired criterion: *maximize the upper deciles*. Consequently, the GenIQ Model allows data analysts to build response and profit models in ways that are not possible with current methods.

The GenIQ Response Model is theoretically superior — with respect to maximizing the upper deciles — to a response model built with alternative response techniques because of the explicit nature of the fitness function. The actual formulation of the fitness function is beyond the scope of this chapter; but suffice to say, the fitness function seeks *to fill the upper deciles with as many responses as possible.*

Alternative response techniques, such as discriminant analysis, logistic regression and artificial neural networks only *implicitly* maximize the desired criterion. Their optimization criteria (fitness function) serves as a surrogate for the desired criterion. Discriminant analysis (DA), with the assumption of bell-shaped data, is defined to explicitly maximize the ratio of between-group sum-of-squares to within-group sum-of-squares.

Logistic regression model (LRM), with the two assumptions of independence of responses and an S-shape relationship between predictor variables and response, is defined to maximize the logistic likelihood function.

Artificial neural network (ANN), a nonparametric method, is typically defined to explicitly minimize mean squared error.

15.8 The GenIQ Profit Model

The GenIQ Profit Model is theoretically superior — with respect to maximizing the upper deciles — to the ordinary least-squares (OLS) regression and ANN. The GenIQ Profit Model uses the genetic methodology with a fitness function that explicitly addresses the desired modeling criterion. This fitness function is defined *to fill the upper deciles with as much profit as possible.* The fitness function for the OLS and ANN models minimizes mean squared error, which serves as a surrogate for the desired criterion.

OLS regression has another weakness in database marketing application. A key assumption of the regression technique is that the dependent variable data must follow a bell-shaped curve. If the assumption is violated, the resultant model may not be accurate and reliable. Unfortunately, profit data are not bell shaped. For example, a 2% response rate yields 98% nonresponders with profit values of zero dollars or some nominal cost associated with nonresponse. Data with a concentration of 98% of a single value cannot be spread out to form a bell-shape distribution.

There is still another data issue when using OLS with database marketing data. Lifetime value (LTV) is an important database marketing performance measure. LTV is typically positively skewed. The log is the appropriate transformation to reshape positively skewed data into a normal curve. However, using the log of LTV as the dependent variable in OLS regression does not guarantee that other OLS assumptions are not violated.[10] Accordingly, attempts at modeling profit with ordinary regression are questionable and/ or difficult.

The GenIQ Response and Profit Models have no restriction on the dependent variable. The GenIQ Models can be accurately and reliably estimated with dependent variables of any shape.[11] This is because the GenIQ estimation is based on the genetic methodolgy, which is inherently nonparametric and assumption-free.

In fact, due to their nonparametric and assumption-free estimation, the GenIQ Models place no restriction on the interrelationship among the independent variables. The GenIQ Models are virtually unaffected by any degree of correlation among the predictor variables. In contrast, OLS and ANN, as well as DA and LRM, can tolerate only a "moderate" degree of correlation among the predictor variables to ensure a stable calculation of their models.

Moreover, the GenIQ Models have no restriction on sample size. GenIQ Models can be built on small samples as well as large samples. OLS and DA and somewhat less for ANN and LRM[12] models require at least a "moderate" size sample.[13]

15.9 Case Study — Response Model

Cataloguer ABC requires a response model based on a recent direct mail campaign, which produced a 0.83% RESPONSE rate. ABC's consultant built a logistic regression model (LRM) using three variables based on the techniques discussed in Chapter 3:

---

[10] The error structure of the OLS equation may not necessarily be normally distributed with zero mean and constant variance, in which case the modeling results are questionable, and additional transformations may be needed.

[11] The dependent variable can be bell-shaped or skewed, bimodal or multimodal, and continuous or discontinuous.

[12] There are specialty algorithms for logistic regression with small sample size.

[13] Statisticians do not agree on how "moderate" a moderate sample size is. I drew a sample of statisticians to determine the size of a moderate sample; the average was 5,000.

1. RENT_1 — a composite variable measuring the ranges of rental cost[14]

2. ACCT_1 — a composite variable measuring the activity of various financial accounts[15]

3. APP_TOTL — the number of inquires

The logistic response model is defined in Equation (15.1) as:

RESPONSE = −1.9 + 0.19*APP_TOTL − 0.24*RENT_1 − 0.25*ACCTS_1 (15.1)

The LRM response validation decile analysis in Table 15.2 shows the performance of the model over chance (i.e., no model). The decile analysis shows a model with good performance in the upper deciles: Cum Lifts for top, 2, 3 and 4 deciles are 264, 174, 157 and 139, respectively. Note that the model may not be as good as initially believed. There is some degree of unstable performance through the deciles, i.e., number of responses do not decrease steadily through the deciles. This unstable performance, which is characterized by "jumps" in deciles 3, 5, 6 and 8, is probably due to an unknown relationship between the three predictor variables and response and/or an important predictor variable that is not included in this model. However, it should be pointed out that only perfect models have perfect performance through the deciles. Good models have some jumps, albeit minor ones.

I built a GenIQ Response Model based on the same three variables used in the LRM. The GenIQ Response Tree is in Figure 15.18. The validation

TABLE 15.2

LRM Response Decile Analysis

Decile	Number of Individuals	Number of Responses	Decile Response Rate	Cumulative Response Rate	Cum Lift Response
Top	1,740	38	2.20%	2.18%	264
2	1,740	12	0.70%	1.44%	174
3	1,740	18	1.00%	1.30%	157
4	1,740	12	0.70%	1.15%	139
5	1,740	16	0.90%	1.10%	133
6	1,740	20	1.10%	1.11%	134
7	1,740	8	0.50%	1.02%	123
8	1,740	10	0.60%	0.96%	116
9	1,740	6	0.30%	0.89%	108
Bottom	1,740	4	0.20%	0.83%	100
Total	17,400	144	0.83%		

---

[14] Four categories of rental cost: less than $200 per month, $200–$300 per month, $300–$500 per month, and greater than $500 per month.
[15] Financial accounts include bank cards, department store cards, installments loans, etc.

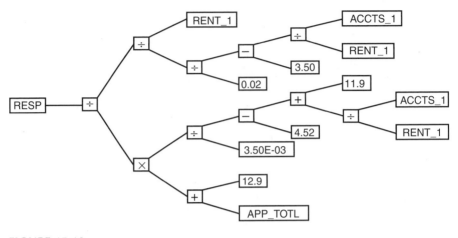

FIGURE 15.18
GenIQ Response Tree

decile analysis, in Table 15.3, shows a model with very good performance in the upper deciles: Cum Lifts for top, 2, 3 and 4 deciles are 306, 215, 167 and 142. In contrast with the LRM, the GenIQ Model has only two minor jumps in deciles 5 and 7. The implication is that the genetic methodology has evolved a better model because its has uncovered a nonlinear relationship among the predictor variables and response. This comparison between LRM and GenIQ is conservative, as GenIQ was assigned and used the same three predictor variables used in LRM. As will be discussed in Chapter 16, GenIQ's strength is in finding its own best set of variables for the prediction task at hand.

The GenIQ Response Model is defined in Equation (15.2) as:

TABLE 15.3

GenIQ Response Decile Analysis

Decile	Number of Individuals	Number of Responses	Decile Response Rate	Cumulative Response Rate	Cum Lift Response
Top	1,740	44	2.50%	2.53%	306
2	1,740	18	1.00%	1.78%	215
3	1,740	10	0.60%	1.38%	167
4	1,740	10	0.60%	1.18%	142
5	1,740	14	0.80%	1.10%	133
6	1,740	10	0.60%	1.02%	123
7	1,740	12	0.70%	0.97%	117
8	1,740	10	0.60%	0.92%	111
9	1,740	8	0.50%	0.87%	105
Bottom	1,740	8	0.50%	0.83%	100
Total	17,400	144	0.83%		

$$\text{RESPONSE} = \frac{0.175 * \text{RENT\_1} * *3}{(\text{ACCTS\_1} - 3.50 * \text{RENT\_1}) * (12.89 + \text{APP\_TOTL}) * (11.92 + \text{ACCTS\_1} - 4.52)}$$

$$(15.2)$$

GenIQ does not outperform LRM across all the deciles in Table 15.4. However, GenIQ yields noticeable Cum Lift improvements for the important first three decile ranges: 16.0%, 23.8% and 6.1%, respectively.

TABLE 15.4

Comparison: LRM and GenIQ Repsonse

Decile	LRM	GenIQ	GenIQ Improvement Over LRM
Top	264	306	16.0%
2	174	215	23.8%
3	157	167	6.1%
4	139	142	2.2%
5	133	133	−0.2%
6	134	123	−8.2%
7	123	117	−4.9%
8	116	111	−4.6%
9	108	105	−2.8%
Bottom	100	100	—

15.10 Case Study — Profit Model

Telecommunications company ATMC seeks to build a zipcode-level model to predict usage, TTLDIAL1. Based on the techniques discussed in Chapter 4, the variables used in building an ordinary regression (OLS) model are:

1. AASSIS_1 — composite of public assistance-related census variables
2. ANNTS_2 — composite of ancestry census variables
3. FEMMAL_2 — composite of gender-related variables
4. FAMINC_1 — a composite variable measuring the ranges of home value[16]

The OLS Profit (Usage) model is defined in Equation (15.3) as:

TTLDIAL1 = 1.5 + −0.35* AASSIS_1 + 1.1*ANNTS_2 + 1.4* FEMMAL_2 +2.8* FAMINC_1

$$(15.3)$$

---

[16] Five categories of home value: less than $100M, $100-$200M, $200-$500M, $500-$750M, greater than $750M

TABLE 15.5

Decile Analysis OLS Profit (Usage) Model

Decile	Number of Customers	Total Dollar Usage	Average Usage	Cumulative Average Usage	Cum Lift Usage
Top	1,800	$38,379	$21.32	$21.32	158
2	1,800	$28,787	$15.99	$18.66	139
3	1,800	$27,852	$15.47	$17.60	131
4	1,800	$24,199	$13.44	$16.56	123
5	1,800	$26,115	$14.51	$16.15	120
6	1,800	$18,347	$10.19	$15.16	113
7	1,800	$20,145	$11.19	$14.59	108
8	1,800	$23,627	$13.13	$14.41	107
9	1,800	$19,525	$10.85	$14.01	104
Bottom	1,800	$15,428	$8.57	$13.47	100
Total	18,000	$242,404	$13.47		

The OLS profit validation decile analysis in Table 15.5 shows the performance of the model over chance (i.e., no model). The decile analysis shows a model with good performance in the upper deciles: Cum Lifts for top, 2, 3 and 4 deciles are 158, 139, 131 and 123, respectively.

I built a GenIQ Profit Model based on the same four variables used in the OLS model. The GenIQ Profit Tree is shown in Figure 15.19. The validation decile analysis in Table 15.6 shows a model with very good performance in the upper deciles: Cum Lifts for top, 2, 3 and 4 deciles are 198, 167, 152 and 140, respectively. This comparison between OLS and GenIQ is conservative, as GenIQ was assigned the same four predictor variables used in OLS. (Curiously, GenIQ only used three of the four variables.) As will be discussed in Chapter 16. GenIQ's strength is finding its own best set of variables for the prediction task at hand.

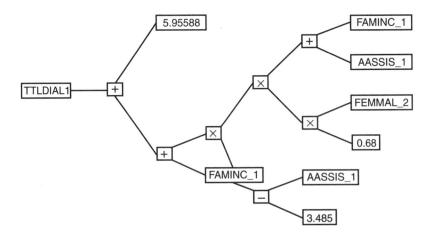

FIGURE 15.19
GenIQ Profit Tree

TABLE 15.6

Decile Analysis GenIQ Profit (Usage) Model

Decile	Number of Customers	Total Dollar Usage	Average Usage	Cumulative Average Usage	Cum Lift Usage
Top	1,800	$48,079	$26.71	$26.71	198
2	1,800	$32,787	$18.22	$22.46	167
3	1,800	$29,852	$16.58	$20.50	152
4	1,800	$25,399	$14.11	$18.91	140
5	1,800	$25,115	$13.95	$17.91	133
6	1,800	$18,447	$10.25	$16.64	124
7	1,800	$16,145	$8.97	$15.54	115
8	1,800	$17,227	$9.57	$14.80	110
9	1,800	$15,125	$8.40	$14.08	105
Bottom	1,800	$14,228	$7.90	$13.47	100
Total	18,000	$242,404	$13.47		

The GenIQ Profit (Usage) Model is defined in Equation (15.4) as:

$$TTLDIAL1 = + 5.95 + FAMINC\_1 + (FAMINC\_1 + AASSIS\_1)*(0.68*FEMMAL\_2)*(AASSIS\_1 - 3.485)) \quad (15.4)$$

GenIQ does outperform OLS across all the deciles in Table 15.7. GenIQ yields noticeable Cum Lift improvements down to the seventh decile: improvements range from 25.5% in the top decile to 6.9% in the 7th decile.

TABLE 15.7

Comparison: OLS and GenIQ Profit (Usage)

Decile	OLS	GenIQ	GenIQ Improvement Over OLS
Top	158	198	25.5%
2	139	167	20.0%
3	131	152	16.2%
4	123	140	14.1%
5	120	133	10.9%
6	113	124	9.3%
7	108	115	6.9%
8	107	110	2.7%
9	104	105	0.6%
Bottom	100	100	—

15.11 Summary

All standard statistical modeling techniques involve optimizing a fitness function to find a specific solution to a problem. The popular ordinary and logistic regression techniques, which seek accurate prediction and classification, respectively, optimize the fitness functions mean squared error (MSE) and the logistic likelihood (LL), respectively. Calculus-based methods are used for the optimization computations.

I presented a new modeling technique, GenIQ Model, which seeks maximum performance (response or profit) from solicitations. The GenIQ Model optimizes the fitness function Cum Lift. GenIQ's optimization computations use the genetic methodology, not the usual calculus. I provided a compendious introduction to genetic methodology with an illustration and inspection of its strengths and limitations.

The GenIQ Model is theoretically superior — with respect to maximizing Cum Lift — to the ordinary and logistic regression models because of its clearly and fully formulated fitness function. The GenIQ fitness function explicitly seeks to fill the upper deciles with as many responses or as much profit as possible. Standard statistical methods only implicitly maximize the Cum Lift, as their fitness functions (MSE and LL) serve as a surrogate maximizing cum lift.

Lastly, I demonstrated the potential of the new technique with response and profit model illustrations. The GenIQ response illustration yields noticeable Cum Lift improvements over logistic regression for the important first three decile ranges: 16.0%, 23.8% and 6.1%, respectively. The GenIQ profit illustration yields noticeable Cum Lift improvements in ordinary regression down to the 7th decile: Improvements range from 25.5% in the top decile to 6.9% in the 7th decile.

References

1. Koza, J., *Genetic Programming: On the Programming of Computers by Means of Natural Selection*, The MIT Press, Cambridge, MA, 1992.

16

Finding the Best Variables for Database Marketing Models[1]

Finding the best possible subset of variables to put in a model has been a frustrating exercise. Many methods of variable selection exist, but none of them is perfect. Moreover, they do not create new variables, which would enhance the predictive power of the original variables themselves. Furthermore, none use a criterion that addresses the specific needs of database marketing models. I present the GenIQ Model as a methodology that uses genetic modeling to find the best variables for database marketing models. Most significantly, genetic modeling can be used to address the specific requirements of database marketing models.

16.1 Background

The problem of finding the best subset of variables in order to define the best model has been extensively studied. Existing methods, based on theory, search heuristics, and rules-of-thumb, each use a unique criterion to build the best model. Selection criteria can be divided into two groups: one based on criteria involving classical hypothesis testing and the other involving residual error sum of squares.[2] Different criteria typically produce different subsets. The number of variables in common with the different subsets is not necessarily large, and the sizes of the subsets can vary considerably.

Essentially, the problem of variable selection is to examine certain subsets and select the subset that either maximizes or minimizes an appropriate criterion. Two subsets are obvious — the best single variable and the complete set of variables. The problem lies in selecting an intermediate subset that is better than both of these extremes. Therefore, the issue is how to find the *necessary variables* among the complete set of variables by deleting both

---

[1] This chapter is based on an article with the same title in *Journal of Targeting, Measurement and Analysis for Marketing*, 9, 3, 2001. Used with permission.

[2] Other criteria are based on information theory and Bayesian rules.

irrelevant variables (variables not affecting the dependent variable) and *redundant variables* (variables not adding anything to the dependent variable). [1]

Reviewed below are five variable selections methods, which are widely used. The first four methods are found in major statistical software packages[3] and the last is the favored rule-of-thumb approach used by many analysts. The test-statistic (TS) for the first three methods uses either the F statistic for a continuous dependent variable, or the G statistic for a binary dependent variable (e.g., response which assumes only two values, yes/no). The TS for the fourth method is either R-squared for a continuous dependent variable, or the Score statistic for a binary dependent variable. The fifth method used the popular corelation coefficient r.

1. *Forward Selection (FS):* This method adds variables to the model until no remaining variable (outside the model) can add anything significant to the dependent variable. FS begins with no variable in the model. For each variable, the TS, a measure of the variable's contribution to the model, is calculated. The variable with the largest TS value that is greater than a preset value C is added to the model. Then the TS is calculated again for the variables still remaining, and the evaluation process is repeated. Thus, variables are added to the model one by one until no remaining variable produces a TS value that is greater than C. Once a variable is in the model, it remains there.

2. *Backward Elimination (BE):* This method deletes variables one by one from the model until all remaining variables contribute something significant to the dependent variable. BE begins with a model which includes all variables. Variables are then deleted from the model one by one until all the variables remaining in the model have TS values greater than C. At each step, the variable showing the smallest contribution to the model (i.e., with the smallest TS value that is less than C) is deleted.

3. *Stepwise (SW):* This method is a modification of the forward selection approach and differs in that variables already in the model do not necessarily stay. As in Forward Selection, SW adds variables to the model one at a time. Variables that have a TS value greater than C are added to the model. After a variable is added, however, SW looks at all the variables already included to delete any variable that does not have a TS value greater than C.

4. *R-squared (R-sq):* This method finds several subsets of different sizes that best predict the dependent variable. R-sq finds subsets of variables that best predict the dependent variable based on the appropriate TS. The best subset of size k has the largest TS value. For a continuous dependent variable, TS is the popular measure

---

[3] SAS/STAT Manual. See PROC REG, and PROC LOGISTIC.

R-squared, the coefficient of multiple determination, which measures the proportion of the "explained" variance in the dependent variable by the multiple regression. For a binary dependent variable, TS is the theoretically correct but less-known Score statistic.[4] R-sq finds the best one-variable model, the best two-variable model, and so forth. However, it is unlikely that one subset will stand out as clearly being the best, as TS values are often bunched together. For example, they are equal in value when rounded at the, say, third place after the decimal point.[5] R-sq generates a number of subsets of each size, which allows the user to select a subset, possibly using nonstatistical conditions.

5. *Rule-of-Thumb Top k Variables (Top-k):* This method selects the top ranked variables in terms of their association with the dependent variable. Each variable's association with the dependent variable is measured by the correlation coefficient r. The variables are ranked by their absolute r values,[6] from largest to smallest. The top-k ranked variables are considered to be the best subset. If the regression model with the top-k variables indicates that each variable is statistically significant, then the set of k variables is declared the best subset. If any variable is not statistically significant, then the variable is removed and replaced by the next ranked variable. The new set of variables is then considered to be best subset, and the evaluation process is repeated.

16.2 Weakness in the Variable Selection Methods

While these methods produce reasonably good models, each method has a drawback specific to its selection criterion. A detailed discussion of the weaknesses is beyond the scope of this chapter; however, there are two common weaknesses that do merit attention. [2,3] *First of all, these methods' selection criteria do not explicitly address the specific needs of database marketing models, namely, to maximize the Cum Lift.*

Second, these methods cannot identify structure in the data. They find the best subset of variables without "digging into the data," a feature which is necessary for finding important variables or structures. Therefore, variable selection methods without "data mining" capability cannot generate the *enhanced best subset.* The following illustration clarifies this weakness. Consider the complete set of variables, $X_1, X_2, ..., X_{10}$. Any of the variable selection methods

---

[4] R-squared theoretically is not the appropriate measure for a binary dependent variable. However, many analysts use it with varying degrees of success.

[5] For example, consider two TS values: 1.934056 and 1.934069. These values are equal when rounding occurs at the third place after the decimal point: 1.934.

[6] Absolute r value means that the sign is ignored. For example, if r = –0.23 then absolute r = +0.23.

in current use will only find the best combination of the original variables (say X_1, X_3, X_7, X_{10}), but can never automatically transform a variable (say, transform X_1 to log X_1) if it were needed to increase the *information content* (*predictive power*) of that variable. Furthermore, none of these methods can generate a re-expression of the original variables (perhaps X_3/X_7) if the constructed variable were to offer more predictive power than the original component variables combined. In other words, current variable selection methods cannot find the *enhanced best subset* that needs to include transformed and re-expressed variables (possibly X_1, X_3, X_7, X_{10}, logX_1, X_3/X_7). A subset of variables without the potential of new variables offering more predictive power clearly limits the analyst in building the best model.

Specifically, these methods fail to identify structure of the types discussed below.

Transformed variables with *a preferred shape.* A variable selection procedure should have the ability to transform an individual variable, if necessary, to induce symmetric distribution. Symmetry is the preferred shape of an individual variable. For example, the workhorse of statistical measures — the mean and variance — is based on the symmetric distribution. Skewed distribution produces inaccurate estimates for means, variances and related statistics, such as the correlation coefficient. Analyses based on skewed distributions typically provide questionable findings. Symmetry facilitates the interpretation of the variable's effect in an analysis. A skewed distribution is difficult to examine because most of the observations are bunched together at one end of the distribution.

A variable selection method also should have the ability to straighten nonlinear relationships. A linear or straight-line relationship is the preferred shape when considering two variables. A straight-line relationship between independent and dependent variables is an assumption of the popular linear model. (Remember that a linear model is defined as a sum of weighted variables, such as $Y = b_0 + b_1{}^*X_1 + b_2{}^*X_2 + b_3{}^*X_3$.)[7] Moreover, straight-line relationships among all the independent variables are a desirable property. [4] Straight-line relationships are easy to interpret: a unit of increase in one variable produces an expected constant increase in a second variable.

Constructed variables from the original variables using simple arithmetic functions. A variable selection method should have the ability to construct simple re-expressions of the the original variables. Sum, difference, ratio, or product variables potentially offer more information than the original variables themselves. For example, when analyzing the efficiency of an automobile engine, two important variables are miles traveled and fuel used (gallons). However, we know the ratio variable of miles per gallon is the best variable for assessing the engine's performance.

Constructed variables from the original variables using a set of functions (e.g., arithmetic, trigonometric, and/or Boolean functions). A variable selection method

---

[7] The weights or coefficients (b_0, b_1, b_2 and b_3) are derived to satisfy some criterion, such as minimize the mean squared error used in ordinary least-square regression, or minimize the joint probability function used in logistic regression.

should have the ability to construct complex re-expressions with mathematical functions that capture the complex relationships in the data, and potentially offer more information than the original variables themselves. In an era of data warehouses and the Internet, big data consisting of hundreds of thousands to millions of individual records and hundreds to thousands of variables are commonplace. Relationships among many variables produced by so many individuals are sure to be complex, beyond the simple straight-line pattern. Discovering the mathematical expressions of these relationships, although difficult without theoretical guidance, should be the hallmark of a high-performance variable selection method. For example, consider the well-known relationship among three variables: the lengths of the three sides of a right triangle. A powerful variable selection procedure would identify the relationship among the sides, even in the presence of measurement error: the longer side (diagonal) is the square root of the sum of squares of the two shorter sides.

In sum, the two above-mentioned weaknesses suggest that a high-performance variable selection method for database marketing models should find the best subset of variables that maximizes the Cum Lift criterion. In the sections that follows, I reintroduce the GenIQ Model of Chapter 15, this time as a high-performance variable selection technique for database marketing models.

16.3 Goals of Modeling in Database Marketing

Database marketers typically attempt to improve the effectiveness of their solicitations by targeting their best customers or prospects. They use a model to identify individuals who are likely to respond to or generate profit[8] from a solicitation. The model provides, for each individual, estimates of probability of response and estimates of contribution-to-profit. Although the precision of these estimates is important, the model's performance is measured at an aggregated level as reported in a decile analysis.

Database marketers have defined the *Cum Lift*, which is found in the decile analysis, as the relevant measure of model performance. Based on the model's selection of individuals, database marketers create a solicitation list to obtain an advantage over a random selection of individuals. The response Cum Lift is an index of how many more responses are expected with a selection based on a model over the expected responses with a random selection (without a model). Similarly, the profit Cum Lift is an index of how much more profit is expected with a selection based on a model over the expected profit with a random selection (without a model). The concept of

---

[8] I use the term "profit" as a stand-in for any measure of an individual's worth, such as sales per order, lifetime sales, revenue, number of visits or number of purchases.

Cum Lift and the steps of its the construction in a decile analysis are presented in Chapter 12.

It should be clear at this point that a model that produces a decile analysis with more responses or profit in the upper (top, 2, 3, or 4) deciles is a better model than a model with less responses or profit in the upper deciles. This concept is the motivation for the GenIQ Model. The GenIQ approach to modeling is to specifically address the objectives concerning database marketers, namely, maximizing response and profit from solicitations. The GenIQ Model uses the genetic methodology to explicitly optimize the desired criterion: maximize the upper deciles. Consequently, the GenIQ Model allows data analysts to build response and profit models in ways that are not possible with current methods.

The GenIQ Response and Profit Models are theoretically superior — with respect to maximizing the upper deciles — to response and profit models built with alternative techniques because of the explicit nature of the fitness function. The actual formulation of the fitness function is beyond the scope of this chapter; suffice to say, the fitness function seeks *to fill the upper deciles with as many responses or as much profit as possible.*

Due to the explicit nature of its fitness criterion and the way it evolves models, the GenIQ Model offers high-performance variable selection for database marketing models. This will become apparent once I illustrate the GenIQ variable selection process in the next section.

16.4 Variable Selection with GenIQ

The best way of explaining variable selection with the GenIQ Model is to illustrate how GenIQ identifies structure in data. In this illustration, I demonstrate finding structure for a response model. GenIQ works equally well for a profit model with a continuous dependent variable.

Cataloguer ABC requires a response model to be built on a recent mail campaign that produced a 3.54% response rate. In addition to the RESPONSE dependent variable, there are nine candidate predictor variables, whose measurements were taken prior to the mail campaign.

1. AGE_Y — Knowledge of customer's age (1 = if known; 0 = if not known)
2. OWN_TEL — Presence of telephone in the household (1 = yes; 0 = no)
3. AVG_ORDE — Average dollar order
4. DOLLAR_2 — Dollars spent within last two years
5. PROD_TYP — Number of different products purchased
6. LSTORD_M — Number of months since last order

7. FSTORD_M — Number of months since first order
8. RFM_CELL — Recency/frequency/money cells (1 = best to 5 = worst)[9]
9. PROMOTION — Number of promotions customer has received

To get an initial read on the information content (predictive power) of the variables, I perform a correlation analysis, which provides the correlation coefficient[10] for each candidate predictor variable with RESPONSE in Table 16.1. The top four variables in descending order of the magnitude[11] of the strength of association are: DOLLAR_2, RFM_CELL, PROD_TYP and LSTORD_M.

I perform five logistic regression analyses (with RESPONSE) corresponding to the five variable selection methods. The resulting best subsets among the nine original variables are represented in Table 16.2. Surprisingly, the forward, backward, and stepwise methods produced the identical subset (DOLLAR_2, RFM_CELL, LSTORD_M, AGE_Y). Because these methods produced a subset size of 4, I set the subset size to 4 for the R-sq and top-k methods. This allows for a fair comparison across all methods. R-sq and top-k produced different best subsets, which include DOLLAR_2, LSTORD_M and AVG_ORDE. It is interesting to note that the most frequently used variables are DOLLAR_2 and LSTORD_M in the Frequency row in Table 16.2. The validation performance of the five logistic models in terms of Cum Lift is reported in Table 16.3. Assessment of model performance at the decile level is as follows:

1. At the top decile, R-sq produced the worst performing model: Cum Lift 239 vs. Cum Lifts 252–256 for the other models.

TABLE 16.1

Correlation Analysis - Nine Original Variable with RESPONSE

Rank	Variable	Corr. Coef. (r)
Top	DOLLAR_2	0.11
2	RFM_CELL	–0.10
3	PROD_TYP	0.08
4	LSTORD_M	–0.07
5	AGE_Y	0.04
6	PROMOTION	0.03
7	AVG_ORDE	0.02
8	OWN_TEL	0.10
9	FSTORD_M	0.01

---

[9] RFM_CELL will be treated as a scalar variable.

[10] I know that the correlation coefficient with or without scatterplots is a crude gauge of predictive power.

[11] The direction of the association is not relevant. That is, the sign of the coefficient is ignored.

TABLE 16.2

Best Subsets among the Nine Original Variables

	DOLLAR_2	RFM_CELL	LSTORD_M	AGE_Y	AVG_ORDE
FS	x	x	x	x	
BE	x	x	x	x	
SW	x	x	x	x	
R-sq	x		x	x	x
Top-4	x	x	x		x
Frequency	5	4	5	4	2

TABLE 16.3

LRM Model Performance Comparison by Variable Selection Methods: Cum Lifts

Decile	FS	BE	SW	R-sq	Top-4	AVG
Top	256	256	256	239	252	252
2	204	204	204	198	202	202
3	174	174	174	178	172	174
4	156	156	156	157	154	156
5	144	144	144	145	142	144
6	132	132	132	131	130	131
7	124	124	124	123	121	123
8	115	115	115	114	113	114
9	107	107	107	107	107	107
Bottom	100	100	100	100	100	100

2. At the second decile, R-sq produced the worst performing model: Cum Lift 198 vs. Cum Lifts 202–204 for the other models.

3. At the third decile, R-sq produced the best performing model: Cum Lift 178 vs. Cum Lifts 172–174 for the other models.

Similar findings can be made at the other depths-of-file.

To facilitate the comparison of the five statistics-based variable selection methods and the proposed genetic-based technique, I use a single measure of model performance for the five methods. The average performance of the five models is measured by AVG, the average of the Cum Lifts across the five methods for each decile, in Table 16.3.

16.4.1 GenIQ Modeling

This section requires an understanding of the genetic methodology and the parameters for controlling a genetic model run (as discussed in Chapter 15).

I set the parameters for controlling the GenIQ Model to run as follows:

1. Population size: 3,000 (models)
2. Number of generations: 250
3. Percentage of the population copied: 10%
4. Percentage of the population used for crossover: 80%
5. Percentage of the population used for mutation: 10%

The GenIQ-variable set consists of the nine candidate predictor variables. For the GenIQ-function set, I select the arithmetic functions (addition, subtraction, multiplication and division), some Boolean operators (and, or, xor, greater/less than), and the log function (Ln). The log function[12] is helpful in symmetrizing typically skewed dollar-amount variables, such as DOLLAR_2. I anticipate that DOLLAR_2 would be part of a genetically-evolved structure defined with the log function. Of course, RESPONSE is the dependent variable.

At the end of the run, 250 generations of copying/crossover/mutation have evolved 750 thousand (250 times 3,000) models according to fitness-proportionate selection. Each model is evaluated in terms of how well it solves the problem of "filling the upper deciles with responders." Good models having more responders in the upper deciles and are more likely to contribute to the next generation of models; poor models having fewer responders in the upper deciles are less likely to contribute to the next generation of models. Consequently, the last generation consists of 3,000 high-performance models, each with a fitness value indicating how well the model solves the problem. The "top" fitness values, typically the fifty largest values,[13] define a *set of fifty "best" models* with equivalent performance (filling the upper deciles with an equivalently large number of responders).

The set of variables defining one of the "best" models has variables in common with the set of variables defining another "best" model. The "common" variables can be considered for the best subset. *The mean incidence of a variable across the set of best models provides a measure for determining the best subset. The GenIQ-selected best subset of original variables consist of variables with mean incidence greater than 0.75.[14] The variables that meet this cut-off score reflect an honest determination of necessary variables with respect to the criterion of maximizing the deciles.*

Returning to the illustration, GenIQ provides the mean incidence of the nine variables across the set of fifty best models in Table 16.4. Thus, the *GenIQ-selected best subset* consists of five variables: DOLLAR_2, RFM_CELL, PROD_TYP, AGE_Y and LSTORD_M.

---

[12] The log to the base 10 also symmetrizes dollar-amount variables.
[13] Top fitness values typically bunch together with equivalent values. The top fitness values are considered equivalent in that their values are equal when rounded at, say, the third place after the decimal point. Consider two fitness values: 1.934056 and 1.934069. These values are equal when rounding occurs at, say, the third place after the decimal point: 1.934.
[14] The mean incidence cut-off score of 0.75 has been empirically predetermined.

TABLE 16.4

Mean Incidence of Original Variables across
the Set of Fifty Best Models

Variable	Mean Incidence
DOLLAR_2	1.43
RFM_CELL	1.37
PROD_TYP	1.22
AGE_Y	1.11
LSTORD_M	0.84
PROMOTION	0.67
AVG_ORDE	0.37
OWN_TEL	0.11
FSTORD_M	0.09

This genetic-based best subset has four variables in common with the
statistics-based best subsets (DOLLAR_2, RFM_CELL, LSTORD_M, and
AGE_Y). Unlike the statistics-based methods, GenIQ finds value in
PROD_TYP and includes it in its best subset in Table 16.5. It is interesting
to note that the most frequently used variables are DOLLAR_2 and
LSTORD_M in the Frequency row in Table 16.5.

At this point, I can assess the predictive power of the genetic-based and
statistics-based best subsets by comparing logistic regression models with
each subset. However, after identifying GenIQ-evolved structure, I chose to
make a more fruitful comparison.

16.4.2 GenIQ-Structure Identification

Just as in nature, where structure is the consequence of natural selection, as
well as sexual recombination and mutation, the GenIQ Model evolves structure via fitness-proportionate selection (natural selection), crossover (sexual
recombination) and mutation. *GenIQ fitness leads to structure, which is evolved*

TABLE 16.5

Best Subsets among Original Variables: Statistics- and Genetic-Based Variable
Selection Methods

Method	DOLLAR_ 2	RFM_ CELL	LSTORD_ M	AGE_Y	AVG_ ORDE	PROD_TYP
FS	x	x	x	x		
BE	x	x	x	x		
SW	x	x	x	x		
R-sq	x		x	x	x	
Top-4	x	x	x		x	
GenIQ	x	x	x	x		x
Frequency	6	5	6	5	2	1

with respect to the criterion of maximizing the deciles. Important structure is found in the best models, typically, the models with the four largest fitness values.

Continuing with the illustration, GenIQ has evolved several structures, or *GenIQ-constructed variables.* The GenIQ Model, in Figure 16.1, has the largest fitness value and reveals five new variables, NEW_VAR1 through NEW_VAR5. Additional structures are found in the remaining three best models: NEW_VAR6 – NEW_VAR8 (in Figure 16.2), NEW_VAR9 (in Figure 16.3), and NEW_VAR10 (in Figure 16.4).

1. NEW_VAR1 = DOLLAR_2/AGE_Y; if Age_Y = 0
   then NEW_VAR1 = 1
2. NEW_VAR2 = (DOLLAR_2)*NEW_VAR1

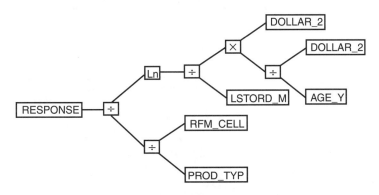

FIGURE 16.1
GenIQ Model, Best #1

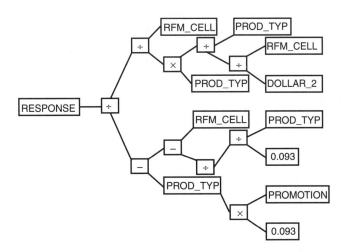

FIGURE 16.2
GenIQ Model, Best #2

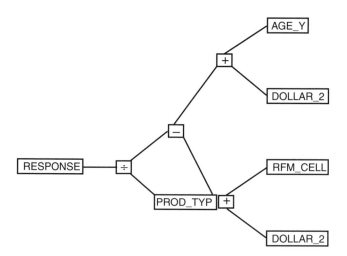

FIGURE 16.3
GenIQ Model, Best #3

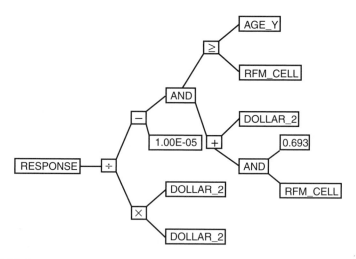

FIGURE 16.4
GenIQ Model, Best #4

3. NEW_VAR3 = NEW_VAR2/LSTORD_M; if LSTORD_M = 0 then NEW_VAR3 = 1

4. NEW_VAR4 = Ln(NEW_VAR3); if NEW_VAR3 not equal to 0 then NEW_VAR4 = 1

5. NEW_VAR5 = RFM_CELL/PROD_TYP; if PROD_TYP = 0 then NEW_VAR5 = 1

6. NEW_VAR6 = RFM_CELL/DOLLAR_2; if DOLLAR_2 = 0 then NEW_VAR6 = 1

7. NEW_VAR7 = PROD_TYP/NEW_VAR6; if NEW_VAR6 = 0 then NEW_VAR7 = 1

8. NEW_VAR8 = NEW_VAR7*PROD_TYP

9. NEW_VAR9 = (AGE_Y/DOLLAR_2) - (RFM_CELL/DOLLAR_2); if DOLLAR_2 = 0 then NEW_VAR9 = 0

10. NEW_VAR10 = 1 if AGE_Y greater than/equal RFM_CELL; otherwise = 0.

To get a read on the predictive power of the new GenIQ-constructed variables, I perform a correlation analysis for each of the nine original variables and the ten new variables with RESPONSE. Some new variables have a stronger association with RESPONSE than the original variables. Specifically, the following associations (larger correlation coefficient) are observed in Table 16.6.

1. NEW_VAR7, NEW_VAR5, NEW_VAR8 and NEW_VAR1 have a stronger association with RESPONSE than the best original variable DOLLAR_2.

2. NEW_VAR10 and NEW_VAR4 fall between the second and third best original variables, RFM_CELL and PROD_TYP.

3. NEW_VAR2 and NEW_VAR3 are ranked eleventh and twelfth in importance before the last two original predictor variables, AGE_Y and PROMOTION.

TABLE 16.6

Correlation Analysis — Nine Original and Ten GenIQ Variables with RESPONSE

Rank	Variable	Corr. Coef. (r)
Top	NEW_VAR7	0.16
2	NEW_VAR5	0.15
3	NEW_VAR8	0.12
4	NEW_VAR1	0.12
5	DOLLAR_2	0.11
6	RFM_CELL	−0.10
7	NEW_VAR10	0.10
8	NEW_VAR4	0.10
9	PROD_TYP	0.08
10	LSTORD_M	−0.07
11	NEW_VAR2	0.07
12	NEW_VAR3	0.06
13	NEW_VAR9	0.05
14	AGE_Y	0.04
15	PROMOTION	0.03
16	NEW_VAR6	−0.02
17	AVG_ORDE	0.02
18	OWN_TEL	0.01
19	FSTORD_M	0.01

16.4.3 GenIQ Variable Selection

The GenIQ-constructed variables plus the GenIQ-selected variables can be thought as an *enhanced best subset* that reflects an honest determination of necessary variables with respect to the criterion of maximizing the deciles. For the illustration data, the enhanced set consists of fifteen variables: DOLLAR_2, RFM_CELL, PROD_TYP, AGE_Y, LSTORD_M and NEW_VAR1 through NEW_VAR10. The predictive power of the enhanced best set is assessed by the comparison between logistic regression models with the genetic-based best subset and with the statistics-based best subset.

Using the enhanced best set, I perform five logistic regression analyses corresponding to the five variable selection methods. The resultant genetic-based best subsets are displayed in Table 16.7. The forward, backward and stepwise methods produced different subsets (of size 4). R-sq(4) and top-4 also produced different subsets. It appears that New_VAR5 is the "most important" variable (i.e., most frequently used variable), as it is selected by all five methods (row Frequency in Table 16.7 equals 5). LSTORD_M is of "second importance," as it is selected by four of the five methods (row Frequency equals 4). RFM_CELL and AGE_Y are the "least important," as they are selected by only one method (row Frequency equals 1).

To assess the gains in predictive power of the genetic-based best subset over the statistics-based best subset, I define AVG-g: the average measure of model validation performance for the five methods for each decile.

Comparison of AVG-g with AVG (average model performance based on the statistics-based set) indicates noticeable gains in predictive power obtained by the GenIQ variable selection technique in Table 16.8. The percentage gains range from an impressive 6.4% (at the sixth decile) to a slight 0.7% (at the ninth decile). The mean percentage gain for the most actionable depth-of-file, the top four deciles, is 3.9%.

This illustration demonstrates the power of the GenIQ variable selection technique over the current statistics-based variable selection methods. GenIQ variable selection is a high-performance method for database marketing models with data mining capability. This method is significant in that it finds the best subset of variables to maximize the Cum Lift criterion.

16.5 Nonlinear Alternative to Logistic Regression Model

The GenIQ Model offers a nonlinear alternative to the logistic regression model (LRM). LRM is a linear approximation of a potentially nonlinear response function, which is typically noisy, multimodal and discontinuous. LRM together with the GenIQ-enhanced best subset of variables provide an unbeatable combination of traditional statistics improved by machine learning. However, this *hybrid GenIQ-LRM Model* is still a linear approximation

TABLE 16.7

Best Subsets among the Enhanced Best Subset Variables

Method	DOLLAR_2	RFM_CELL	PROD_TYP	AGE_Y	LSTORD_M	NEW_VAR1	NEW_VAR4	NEW_VAR5
FS			x		x		x	x
BE	x			x	x		x	x
SW			x		x			x
R-sq			x		x	x		x
Top-4		x				x	x	x
Frequency	1	1	3	1	4	2	3	5

TABLE 16.8

Model Performance Comparison Based on the Genetic-Based Best Subsets: Cum Lifts

Decile	FS	BE	SW	R-sq	Top-4	AVG-g	AVG	Gain
Top	265	260	262	265	267	264	252	4.8%
2	206	204	204	206	204	205	202	1.2%
3	180	180	180	178	180	180	174	3.0%
4	166	167	167	163	166	166	156	6.4%
5	148	149	149	146	149	148	144	3.1%
6	135	137	137	134	136	136	131	3.3%
7	124	125	125	123	125	124	123	1.0%
8	116	117	117	116	117	117	114	1.9%
9	108	108	108	107	108	108	107	0.7%
Bottom	100	100	100	100	100	100	100	0.0%

of a potentially nonlinear response function. The GenIQ Model itself — as defined by the entire tree with all its structure — is a nonlinear super-structure with a strong possibility for further improvement over the hybrid GenIQ-LRM, and, of course, over LRM. Because the degree of nonlinearity in the response function is never known, the best approach is to compare the GenIQ Model with the hybrid GenIQ-LRM model. If the improvement is determined to be stable and noticeable, then the GenIQ Model should be used.

Continuing with the illustration, the GenIQ Model Cum Lifts are reported in Table 16.9. The GenIQ Model offers noticeable improvements over the performance of hybrid GenIQ-LRM Model (AVG-g). The percentage gains (column 5) range from an impressive 7.1% (at the sixth decile) to a respectable 1.2% (at the ninth decile). The mean percentage gain for the most actionable depth-of-file, the top four deciles, is 4.6%.

I determine the improvements of the GenIQ Model over the performance of the LRM (AVG). The mean percentage gain (column 6) for the most actionable depth-of-file, the top four deciles, is 8.6%, which includes a huge 12.2% in the top decile.

Note that a set of four separate GenIQ Models is needed to obtain the reported decile performance levels because no single GenIQ Model could be evolved to provide gains *for all upper deciles*. The GenIQ Models that produce the top, second, third and fourth deciles are in Figures 16.1, 16.5, 16.6 and 16.7, respectively. The GenIQ Model that produced the fifth through bottom deciles is in Figure 16.8.

A set of GenIQ Models is required when the response function is nonlinear with noise, multi-peaks and discontinuities. The capability of GenIQ to generate many models with desired performance gains reflects the flexibility of the GenIQ paradigm. It allows for intelligent and adaptive modeling of the data to account for the variation of an apparent nonlinear response function.

TABLE 16.9

Model Performance Comparison LRM and GenIQ Model: Cum Lifts

Decile	AVG-g (Hybrid)	AVG (LRM)	GenIQ	GenIQ Gain Over	
				Hybrid	LRM
Top	264	252	283	7.1%	12.2%
2	205	202	214	4.4%	5.6%
3	180	174	187	3.9%	7.0%
4	166	156	171	2.9%	9.5%
5	148	144	152	2.8%	5.9%
6	136	131	139	2.5%	5.9%
7	124	123	127	2.2%	3.2%
8	117	114	118	1.3%	3.3%
9	108	107	109	1.2%	1.9%
Bottom	100	100	100	0.0%	0.0%

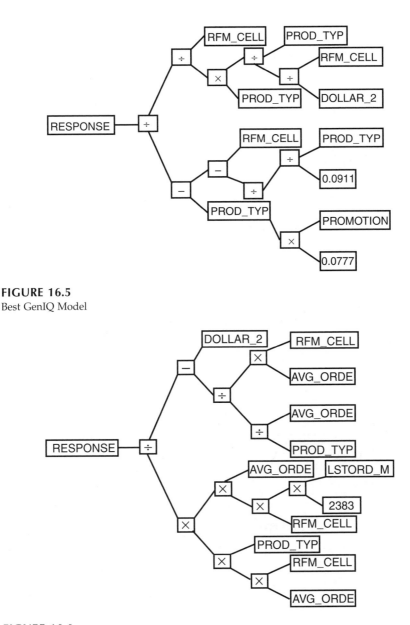

FIGURE 16.5
Best GenIQ Model

FIGURE 16.6
Best GenIQ Model

This illustration shows the power of the GenIQ Model as a nonlinear alternative to the logistic regression model. GenIQ provides a two-step procedure for response modeling. First, build the best hybrid GenIQ-LRM model. Second, select the best GenIQ Model. If the GenIQ Model offers a stable and noticeable improvement over the hybrid model, then the GenIQ Model is the preferred response model.

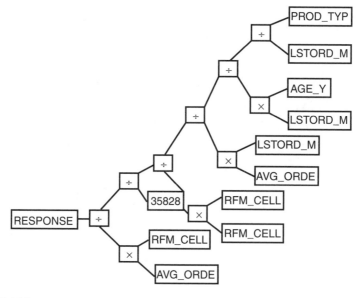

FIGURE 16.7
Best GenIQ Model

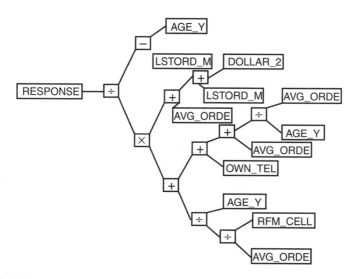

FIGURE 16.8
Best GenIQ Model

As previously mentioned, the GenIQ Model works equally well for finding structure in a profit model. Accordingly, GenIQ Model is a nonlinear alternative to the ordinary least-squares regression model (OLS). The GenIQ Model offers potentially stable and noticeable improvement over OLS, and the hybrid GenIQ-OLS Model.

16.6 Summary

After framing the problem of variable selection with the five popular statistics-based methods, I pointed out two common weaknesses of the methods. Each hinder its capacity to achieve the desired requirement of database marketing model: neither identifying structure, nor explicitly maximizing the Cum Lift criterion.

I have presented the GenIQ Model as a genetic-based approach for variable selection for database marketing models. The GenIQ Response and Profit Models are theoretically superior — with respect to maximizing the upper deciles — to response and profit models built with logistic and ordinary regression models, respectively, because of the nature of its fitness function. The GenIQ fitness function explicitly seeks to fill the upper deciles with as many responses or as much profit as possible. Standard statistical methods only implicitly maximize the Cum Lift, as their fitness functions serve as a surrogate maximizing Cum Lift.

Using a response model illustration, I demonstrated the GenIQ Model as a high-performance variable selection method with data mining capability for finding important structure to maximize the Cum Lift criterion. Starting with nine candidate predictor variables, the statistics-based variable selection methods identified five predictor variables in defining its best subsets. GenIQ also identified five predictor variables, of which four were in common with the statistics-based best subsets. In addition, GenIQ evolved ten structures (new variables), of which four had a stronger association with response than the best original predictor variable. Two new variables fell between the second and third best original predictor variables. As a result, GenIQ created the enhanced best subset of fifteen variables.

The GenIQ variable selection method outperformed the statistics-based variable selection methods. I built logistic regression models (LRM) for the five statistics-based variable selection methods using the enhanced best subset and compared its "average" performance (AVG-g) with the "average" performance (AVG) of the LRM for the five statistics-based methods using the original nine variables. Comparison of AVG-g with AVG indicates noticeable gains in predictive power: the percentage gains range from an impressive 6.4% to a slight 0.7%. The mean percentage gain for the most actionable depth-of-file, the top four deciles, is 3.9%.

Last, I advanced the GenIQ Model itself as a nonlinear alternative to the standard regression models. LRM together with the GenIQ-enhanced best subset of variables provide an unbeatable combination of traditional statistics improved by machine learning. However, this hybrid GenIQ-LRM Model is still a linear approximation of a potentially nonlinear response function. The GenIQ Model itself — as defined by the entire tree with all its structure — is a nonlinear superstructure with a strong possibility for further improvement over the hybrid GenIQ-LRM. For the response illustration, the set of GenIQ Models produced noticeable improvements over the performance of hybrid GenIQ-LRM Model. The percentage gains range from an impressive 7.1% to a respectable 1.2%. The mean percentage gain for the most actionable depth-of-file, the top four deciles, is 4.6%.

References

1. Dash, M. and Liu, H., Feature selection for classification, *Intelligent Data Analysis*, Elsevier Science, New York, 1997.
2. Ryan, T.P., *Modern Regression Methods*, Wiley, New York, 1997.
3. Miller, A.J., *Subset Selection in Regression*, Chapman and Hall, London, 1990.
4. Fox, J., *Applied Regression Analysis, Linear Models, and Related Methods*, Sage Publications, Thousand Oaks, CA, 1997.

17

Interpretation of Coefficient-Free Models

The ordinary regression model is the thought of reference when database marketers hear the words "new kind of model." Data analysts use the regression concept and its prominent characteristics when judiciously evaluating an alternative modeling technique. This is because the ordinary regression paradigm is the underpinning for the solution to the ubiquitous prediction problem. Marketers with limited statistical background undoubtedly draw on their educated notions of the regression model before accepting a new technique. New modeling techniques are evaluated by the coefficients they produce. If the new coefficients impart comparable information to the prominent characteristic of the regression model — the regression coefficient — then the new technique passes the first line of acceptance. If not, the technique is summarily rejected. A quandary arises when a new modeling technique, like some machine learning methods, produces models with no coefficients. The primary purpose of this chapter is to present a method for calculating a quasi-regression coefficient, which provides a frame of reference for evaluating and using coefficient-free models. Secondarily, the quasi-regression coefficient serves as a trusty assumption-free alternative to the regression coefficient, which is based on an implicit and hardly-tested assumption necessary for reliable interpretation.

17.1 The Linear Regression Coefficient

The redoubtable regression coefficient, formally known as the ordinary linear regression coefficient, linear-RC(ord), enjoys everyday use in database marketing analysis and modeling. The common definition of the *linear-RC(ord) for predictor variable X* is *the predicted (expected) constant change in the dependent variable Y associated with a unit-change in X*. The usual mathematical expression for the coefficient is in Equation (17.1). Although correctly stated, the definition requires commentary to have thorough understanding for expertness in its use. I parse the definition of linear-RC(ord), and then provide two illustrations of the statistical measure within the openwork of the linear regression paradigm, which serve as a backdrop for presenting the quasi-regression coefficient.

$$\text{Linear-RC(ord)} = \frac{\text{Predicted-Change in Y}}{\text{Unit-Change in X}} \tag{17.1}$$

Consider the simple ordinary linear regression model of Y on X based on a sample of (X_i, Y_i) points: pred_Y = a + b*X.

1. *Simple* means one predictor variable X is used.
2. *Ordinary* connotes that the dependent variable Y is continuous.
3. *Linear* has a dual meaning. Explicitly, linear denotes that the model is defined by the sum of the weighted predictor variable b*X and the constant *a*. Implicitly, it connotes the linearity assumption that the true relationship between Y and X is straight-line.
4. *Unit-change in X* means the difference of value 1 for two X values ranked in ascending order, X_r and X_{r+1}, i.e., $X_{r+1} - X_r = 1$.
5. *Change in Y* means the difference in predicted Y, pred_Y, values corresponding to $(X_r,\ \text{pred\_Y}_r)$ and $(X_{r+1},\ \text{pred\_Y}_{r+1})$: $\text{pred\_Y}_{r+1} - \text{pred\_Y}_r$.
6. *Linear-RC(ord)* indicates that the expected change in Y is constant, namely, b.

17.1.1 Illustration for the Simple Ordinary Regression Model

Consider the simple ordinary linear regression of Y on X based on the ten observations in the dataset A in Table 17.1. The satisfaction of the linearity assumption is indicated by the plot of Y vs. X; there is an observed positive straight-line relationship between the two variables. The X-Y plot and the

TABLE 17.1

Dataset A

Y	X	pred_Y
86	78	84.3688
74	62	71.6214
66	58	68.4346
65	53	64.4511
64	51	62.8576
62	49	61.2642
61	48	60.4675
53	47	59.6708
52	38	52.5004
40	19	37.3630

plot of residual vs. predicted Y both suggest that the resultant regression model in Equation (17.2) is reliable. (These types of plots are discussed in Chapter 3. The plots for these data are not shown.) Accordingly, the model provides a level of assurance that the estimated linear-RC(ord) value of 0.7967 is a reliable point estimator of the true linear regression coefficient of X. Thus, for each and every unit-change in X, between observed X values of 19 and 78, the expected constant change in Y is 0.7967.

$$Pred\_Y = 22.2256 + 0.7967*X \qquad (17.2)$$

17.1.2 Illustration for the Simple Logistic Regression Model

Consider the simple logistic regression of response Y on X based on the ten observations in dataset B in Table 17.2. Recall that the logistic regression model predicts the logit Y, and is *linear* in the same ways as the ordinary regression model. It is a linear model, as it is defined by the sum of a weighted predictor variable plus a constant, and has the linearity assumption that the underlying relationship between the logit Y and X is straight-line. Accordingly, the definition of the simple logistic linear regression coefficient, *linear-RC(logit) for predictor variable X is the expected constant change in the logit Y associated with a unit-change in X.*

The smooth plot of logit Y vs. X is inconclusive as to the linearity between the logit Y and X, undoubtedly due to only ten observations. However, the plot of residual vs. predicted logit Y suggests that the resultant regression model in Equation (17.3) is reliable. (The smooth plot is discussed in Chapter 3. Plots for these data are not shown.) Accordingly, the model provides a level of assurance that the estimated linear-RC(logit) value of 0.1135 is a reliable point estimator of the true linear logistic regression coefficient of X.

TABLE 17.2

Dataset B

Y	X	pred_lgt Y	pred_prb Y
1	45	1.4163	0.8048
1	35	0.2811	0.5698
1	31	−0.1729	0.4569
1	32	−0.0594	0.4851
1	60	3.1191	0.9577
0	46	1.5298	0.8220
0	30	−0.2865	0.4289
0	23	−1.0811	0.2533
0	16	−1.8757	0.1329
0	12	−2.3298	0.0887

Thus, for each and every unit-change in X, between observed X values of 12 and 60, the expected constant change in the logit Y is 0.1135.

$$\text{Pred\_Logit } Y = -3.6920 + 0.1135 * X \qquad (17.3)$$

17.2 The Quasi-Regression Coefficient for Simple Regression Models

I present the quasi-regression coefficient, quasi-RC, for the simple regression model of Y on X. The *quasi-RC for predictor variable X is the expected change — not necessarily constant — in dependent variable Y per unit-change in X.* The distinctness of the quasi-RC is that it offers a generalization of the linear-RC. It has a flexibility to measure nonlinear relationships between the dependent and predictor variables. I outline the method for calculating the quasi-RC and motivate its utility by applying the quasi-RC to the ordinary regression illustration above. Then I continue the usance of the quasi-RC method for the logistic regression illustration, showing how it works for linear and nonlinear predictions.

17.2.1 Illustration of Quasi-RC for the Simple Ordinary Regression Model

Continuing with the simple ordinary regression illustration, I outline the steps for deriving the quasi-RC(ord) in Table 17.3:

1. Score the data to obtain the predicted Y, pred_Y. (Column 3, Table 17.1.)
2. Rank the data in ascending order by X, and form the pair (X_r, X_{r+1}). (Columns 1 and 2, Table 17.3.)
3. Calculate the change in X: $X_{r+1} - X_r$. (Column 3 = column 2 – column 1, Table 17.3.)
4. Calculate the change in predicted Y: $\text{pred\_Y}_{r+1} - \text{pred\_Y}_r$. (Column 6 = column 5 – column 4, Table 17.3.)
5. Calculate the quasi-RC(ord) for X: change in predicted Y divided by change in X. (Column 7 = column 6/column 3, Table 17.3.)

The quasi-RC(ord) is constant across the nine (X_r, X_{r+1}) intervals and equals the estimated linear RC(ord) value of 0.7967. In superfluity, it is constant for each and every unit-change in X within each of the X intervals: between 19 and 38, 38 and 47, ..., and 62 and 78. This is no surprise as the predictions are from a linear model. Also, the plot of pred_Y vs. X (not shown) shows a perfectly positively sloped straight-line, whose slope is 0.7967. (Why?)

TABLE 17.3

Calculations for Quasi-RC(ord)

X_r	X_r+1	change_X	pred_Y_r	pred_Y_r+1	change_Y	quasi-RC(ord)
19	19	.	.	37.3630	.	.
38	38	19	37.3630	52.5005	15.1374	0.7967
47	47	9	52.5005	59.6709	7.1704	0.7967
48	48	1	59.6709	60.4676	0.7967	0.7967
49	49	1	60.4676	61.2643	0.7967	0.7967
51	51	2	61.2643	62.8577	1.5934	0.7967
53	53	2	62.8577	64.4511	1.5934	0.7967
58	58	5	64.4511	68.4346	3.9835	0.7967
62	62	4	68.4346	71.6215	3.1868	0.7967
	78	16	71.6215	84.3688	12.7473	0.7967

17.2.2 Illustration of Quasi-RC for the Simple Logistic Regression Model

By way of further motivation for the quasi-RC methodology, I apply the five steps to the logistic regression illustration with the appropriate changes to accommodate working with logit units for deriving the quasi-RC(logit) in Table 14.4:

1. Score the data to obtain the predicted logit Y, pred_lgt Y. (Column 3, Table 17.2.)
2. Rank the data in ascending order by X and form the pair (X_r, X_{r+1}). (Columns 1 and 2, Table 17.4.)
3. Calculate the change in X: $X_{r+1} - X_r$. (Column 3 = column 2 – column 1, Table 17.4.)
4. Calculate the change in predicted logit Y: pred_lgt Y_{r+1} – pred_lgt Y_r. (Column 6 = column 5 – column 4, Table 17.4.)
5. Calculate the quasi-RC for X (logit): change in predicted logit Y divided by change in X. (Column 7 = column 6/column 3, Table 17.4.)

The quasi-RC(logit) is constant across the nine (X_r, X_{r+1}) intervals and equals the estimated linear-RC(logit) value of 0.1135. Again, in superfluity, it is constant for each and every unit-change within the X intervals: between 12 and 16, 16 and 23, ..., and 46 and 60. Again, this is no surprise as the predictions are from a linear model. Also, the plot of predicted logit Y vs. X (not shown) shows a perfectly positively sloped straight line, whose slope is 0.1135. Thus far, the two illustrations show how the quasi-RC method works and holds up, for simple linear predictions, i.e., predictions produced by linear models with one predictor variable.

In the next section, I show how the quasi-RC method works with a simple nonlinear model in its efforts to provide an honest attempt at imparting regression coefficient-like information. A nonlinear model is defined, in earnest, as a model that is not a linear model, i.e., is not a sum of weighted predictor variables. The simplest nonlinear model — the probability of response — is a restatement of the simple logistic regression of response Y on X, defined in Equation (17.4).

$$\text{Probability of response } Y = \exp(\text{logit } Y)/(1 + \exp(\text{logit } Y)) \quad (17.4)$$

Clearly, this model is nonlinear. It is said to be nonlinear in its predictor variable X, which means that the expected change in the probability of response *varies* as the unit-change in X varies through the range of observed X values. Accordingly, the *quasi-RC(prob) for predictor variable X is the expected change — not necessarily constant — in the probability Y per unit-change in the X.* In the next section, I conveniently use the logistic regression illustration above to show how the quasi-RC method works with nonlinear predictions.

TABLE 17.4

Calculations for Quasi-RC(logit)

X_r	X_r+1	change_X	pred_lgt_r	pred_lgt_r+1	change_lgt	quasi-RC(logit)
12	12	.	.	-2.3298	.	.
16	16	4	-2.3298	-1.8757	0.4541	0.1135
23	23	7	-1.8757	-1.0811	0.7946	0.1135
30	30	7	-1.0811	-0.2865	0.7946	0.1135
31	31	1	-0.2865	-0.1729	0.1135	0.1135
32	32	1	-0.1729	-0.0594	0.1135	0.1135
35	35	3	-0.0594	0.2811	0.3406	0.1135
45	45	10	0.2811	1.4163	1.1352	0.1135
46	46	1	1.4163	1.5298	0.1135	0.1135
	60	14	1.5298	3.1191	1.5893	0.1135

TABLE 17.5

Calculations for Quasi-RC(prob)

X_r	X_r+1	change_X	prob_Y_r	prob_Y_r+1	change_prob	quasi-RC(prob)
	12	.	.	0.0887	.	.
12	16	4	0.0887	0.1329	0.0442	0.0110
16	23	7	0.1329	0.2533	0.1204	0.0172
23	30	7	0.2533	0.4289	0.1756	0.0251
30	31	1	0.4289	0.4569	0.0280	0.0280
31	32	1	0.4569	0.4851	0.0283	0.0283
32	35	3	0.4851	0.5698	0.0847	0.0282
35	45	10	0.5698	0.8048	0.2349	0.0235
45	46	1	0.8048	0.8220	0.0172	0.0172
46	60	14	0.8220	0.9577	0.1357	0.0097

17.2.3 Illustration of Quasi-RC for Nonlinear Predictions

Continuing with the logistic regression illustration, I outline the steps for deriving the quasi-RC(prob) in Table 17.5 by straightforwardly modifying the steps for the quasi-RC(logit) to account for working in probabilty units.

1. Score the data to obtain the predicted logit of Y, pred_lgt Y. (Column 3, Table 17.2.)
2. Convert the pred_lgt Y to predicted probability Y, pred_prb Y. (Column 4, Table 17.2.) The conversion formula is: probability Y equals exp(logit Y) divided by the sum of 1 plus exp(logit Y).
3. Rank the data in ascending order by X, and form the pair (X_r, X_{r+1}). (Columns 1 and 2, Table 17.5.)
4. Calculate change in X: $X_{r+1} - X_r$. (Column 3 = column 2 – column 1, Table 17.5.)
5. Calculate change in probability Y: pred_prb Y_{r+1} – pred_prb Y_r. (Column 6 = column 5 – column 4, Table 17.5.)
6. Calculate the quasi-RC(prob) for X: change in probability Y divided by the change in X. (Column 7 = column 6/column 3, Table 17.5.)

Quasi-RC(prob) varies as X — in a nonlinear manner — goes through its range between 12 and 60. The quasi-RC values for the nine intervals are 0.0110, 0.0172, …, 0.0097, respectively, in Table 17.5. This is no surprise as the general relationship between probability of response and a given predictor variable has a theoretical prescribed nonlinear S-shape (known as an ogive curve). The plot of probability of Y vs. X in Figure 17.1 reveals this nonlinearity, although the limiting ten points may make it too difficult to see.

Thus far, the three illustrations show how the quasi-RC method works and holds up for linear and nonlinear predictions based on the simple one-predictor variable regression model. In the next section, I extend the method beyond the simple one-predictor variable regression model to *the everymodel*, any multiple linear or nonlinear regression model, or any coefficient-free model.

17.3 Partial Quasi-RC for the Everymodel

The interpretation of the regression coefficient in the multiple (two or more predictor variables) regression model essentially remains the same as its meaning in the simple regression model. The regression coefficient is formally called the *partial* linear regression coefficient, partial linear-RC, which connotes that the model has other variables, whose effects are partialled out of the relationship between the dependent variable and the predictor variable

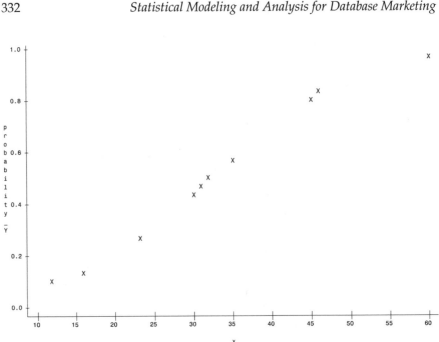

FIGURE 17.1
Plot of Probability of Y vs. X

under consideration. The *partial linear-RC for predictor variable X is the expected constant change in the dependent variable Y associated with a unit-change in the X when the other variables are held constant.* This is the well-accepted interpretation of the partial linear regression coefficient (as discussed in Chapter 6, Section 6.6).

The reading of the partial linear-RC for a given predictor variable is based on an implicit assumption that the statistical adjustment — which removes the effects of the other variables from the dependent variable and the predictor variable — produces a linear relationship between the dependent variable and the predictor variable. Although the workings of statistical adjustment are theoretically sound, it does not guarantee linearity between the adjusted-dependent and the adjusted-predictor variable. In general, an assumption based on the property of linearity is tenable. In the present case of statistical adjustment, the likelihood of the linearity assumption holding tends to decrease as the number of other variables increases. Interesting, it is not a customary effort to check the validity of the linearity assumption, which could render the partial linear-RC questionable.

The quasi-RC method provides the *partial quasi-RC* as a trusty assumption-free alternative measure of the "expected change in the dependent variable" without reliance on statistical adjustment, and restriction of a linear relationship between dependent variable and the predictor variable. Formally, the *partial quasi-RC for predictor variable X is the expected change — not necessarily constant — in the dependent variable Y associated with a unit-change in X when the other variables are held constant.* The quasi-RC method provides a flexibility, which enables the data analyst to:

1. Validate an overall linear trend in the dependent variable vs. the predictor variable for given values within the other-variables region (i.e., given the other variables are held constant). For linear regression models, the method serves as a diagnostic to test the linearity assumption of the partial linear-RC. If the test result is positive (i.e., a nonlinear pattern emerges), which is a symptom of an incorrect structural form of the predictor variable, then a remedy can be inferred (i.e., a choice of re-expression of the predictor variable to induce linearity with the dependent variable).

2. Consider the liberal view of a nonlinear pattern in the dependent variable vs. the predictor variable for given values within the other-variables region. For nonlinear regression models, the method provides an EDA procedure to uncover underlying structure of the "expected change in the dependent variable."

3. Obtain coefficient-like information from coefficient-free models. This information encourages the use of "black box" machine learning methods, which are characterized by the absence of regression-like coefficients.

In the next section I outline the steps for calculating the partial quasi-RC for the everymodel. I provide an illustration using a multiple logistic regression model to show how the method works and how to interpret the results. In the last section of this chapter, I apply the partial quasi-RC method to the coefficient-free GenIQ Model presented in Chapter 15.

17.3.1 Calculating the Partial Quasi-RC for the Everymodel

Consider the everymodel for predicting Y based on four predictor variables X_1, X_2, X_3 and X_4. The calculations and guidelines for the partial quasi-RC for X_1 are as follows.

1. To affect the "holding constant" of the other variables $\{X_2, X_3, X_4\}$, consider the typical values of the *M-spread common region*. For example, the M20-spread common region consists of the individuals whose $\{X_2, X_3, X_4\}$-values are common to the individual M20-spreads (the middle 20% of the values) for each of the other variables; i.e., common to M20-spread for X_2 and M20-spread for X_3 and M20-spread for X_4. Similarly, the M25-spread common region consists of the individuals whose $\{X_2, X_3, X_4\}$ values are common to the individual M25-spreads (the middle 25% of the values) for each of the other variables.

2. The size of the common region is clearly based on the number and measurement of the other variables. A rule of thumb for sizing the region for reliable results is as follows. The initial M-spread common region is M20. If partial quasi-RC values seem suspect, then increase the common region by 5%, resulting in a M25-spread region. Increase the common region by 5% increments until the partial quasi-RC results are trustworthy. Note: A 5% increase is a nominal 5% because increasing each of the other-variables M-spread by 5% does not necessarily increase the common region by 5%.

3. For any of the other variables, whose measurement is coarse (includes a handful of distinct values), its individual M-spread may need to be decreased by 5% intervals until the partial quasi-RC values are trustworthy.

4. Score the data to obtain the predicted Y for all individuals in the common M-spread region.

5. Rank the scored data in ascending order by X_1.

6. Divide the data into equal-sized slices by X_1. In general, if the expected relationship is linear as when working with a linear model and testing the partial linear-RC linearity assumption, then start with five slices; increase the number of slices as required to obtain a trusty relationship. If the expected relationship is nonlinear as when working with a nonlinear regression model, then start with ten slices; increase the number of slices as required to obtain a trusty relationship.

7. The number of slices depends on two matters of consideration, the size of the M-spread common region, and the measurement of the predictor variable for which partial quasi-RC is being derived. If the common region is small, then a large number of slices tends to produce unreliable quasi-RC values. If the region is large, then a large number of slices, which otherwise does not pose a reliability concern, may produce untenable results, as offering too liberal a view of the overall pattern. If the measurement of the predictor variable is coarse, the number of slices equals to the number of distinct values.

8. Calculate the minimum, maximum and median of X_1 within each slice, form the pair (median $X_{\text{slice } i}$, median $X_{\text{slice } i+1}$).

9. Calculate the change in X_1: median $X_{\text{slice } i+1}$ − median $X_{\text{slice } i}$.

10. Calculate the median of the predicted Y within each slice, and form the pair (median pred_$Y_{\text{slice } i}$, median $Y_{\text{slice } i+1}$).

11. Calculate the change in predicted Y: median pred_$Y_{\text{slice } i+1}$ − median pred_$Y_{\text{slice } i}$.

12. Calculate the partial quasi-RC for X_1: change in predicted Y divided by change in X_1.

17.3.2 Illustration for the Multiple Logistic Regression Model

Consider the illustration in Chapter 16 of cataloger ABC, who requires a response model built on a recent mail campaign. I build a logistic regression model (LRM) for predicting RESPONSE based on four predictor variables:

1. DOLLAR_2 — Dollars spent within last two years
2. LSTORD_M — Number of months since last order
3. RFM_CELL — Recency/frequency/money cells (1 = best to 5 = worst)
4. AGE_Y — Knowledge of customer's age (1 = if known; 0 = if not known)

The RESPONSE model in Equation (17.4):

$$\text{Pred\_lgt RESPONSE} = -3.004 + 0.00210 * \text{DOLLAR\_2}$$

$$- 0.1995 * \text{RFM\_CELL} - 0.0798 * \text{LSTORD\_M} + 0.5337 * \text{AGE\_Y} \quad (17.4)$$

I detail the calculation for deriving the LRM partial quasi-RC(logit) for DOLLAR_2 in Table 17.6.

1. Score the data to obtain Pred_lgt RESPONSE for all individuals in the M-spread common region.
2. Rank the scored data in ascending order by DOLLAR_2, and divide the data into five slices, in column 1, by DOLLAR_2.
3. Calculate the minimum, maximum and median of DOLLAR_2, in columns 2, 3 and 4, respectively, for each slice, and form the pair (median DOLLAR_2 $_{\text{slice } i}$, median DOLLAR_2 $_{\text{slice } i+1}$) in columns 4 and 5, respectively.
4. Calculate the change in DOLLAR_2: median DOLLAR_2$_{\text{slice } i+1}$ − median DOLLAR_2 $_{\text{slice } i}$ (column 6 = column 5 − column 4).
5. Calculate the median of the predicted logit RESPONSE within each slice, and form the pair (median Pred_lgt RESPONSE $_{\text{slice } i}$, median Pred_lgt RESPONSE $_{\text{slice } i+1}$) in columns 7 and 8.
6. Calculate the change in Pred_lgt RESPONSE: median Pred_lgt RESPONSE $_{\text{slice } i+1}$ − median Pred_lgt RESPONSE $_{\text{slice } i}$ (column 9 = column 8 − column 7).
7. Calculate the partial quasi-RC(logit) for DOLLAR_2: the change in the Pred_lgt RESPONSE divided by the change in DOLLAR_2, for each slice (column 10 = column 9/column 6.)

TABLE 17.6

Calculations for LRM Partial Quasi-RC(logit): DOLLAR_2

Slice	min_DOLLAR_2	max_DOLLAR_2	med_DOLLAR_2	med_DOLLAR_2_r	med_DOLLAR_2_r+1	change_DOLLAR_2	med_lgt_r	med_lgt_r+1	change_lgt	quasi-RC (logit)
1	0	43	.	.	40	.	.	-3.5396	.	.
2	43	66	40	40	50	10	-3.5396	-3.5276	0.0120	0.0012
3	66	99	50	50	80	30	-3.5276	-3.4810	0.0467	0.0016
4	99	165	80	80	126	46	-3.4810	-3.3960	0.0850	0.0018
5	165	1,293	126	126	242	116	-3.3960	-3.2219	0.1740	0.0015

The LRM partial quasi-RC(logit) for DOLLAR_2 is interpreted as follows:

1. For slice 2, which has minimum and maximum DOLLAR_2 values of 43 and 66, respectively, the partial quasi-RC(logit) is 0.0012. This means that for each and every unit-change in DOLLAR_2 between 43 and 66, the expected constant change in the logit RESPONSE is 0.0012.

2. Similarly, for slices 3, 4 and 5, the expected constant changes in the logit RESPONSE within the corresponding intervals are 0.0016, 0.0018 and 0.0015, respectively. Note that for slice 5, the maximum DOLLAR_2 value, in column 3, is 1293.

3. At this point, the pending implication is that there are four levels of expected change in the logit RESPONSE associated by DOLLAR_2 across its range from 43 to 1293.

4. However, the partial quasi-RC plot for DOLLAR_2 of the relationship of the smooth predicted logit RESPONSE (column 8) vs. the smooth DOLLAR_2 (column 5), in Figure 17.2, indicates there is a single expected constant change across the DOLLAR_2 range, as the variation among slice-level changes is reasonably due to sample variation. This last examination supports the decided implication

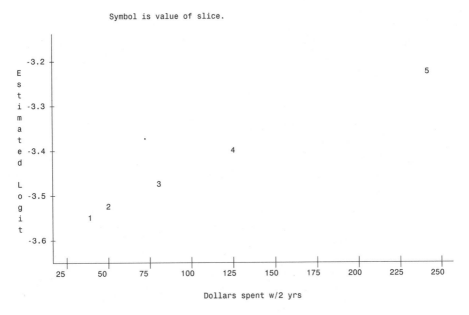

FIGURE 17.2
Visual Display of LRM Partial Quasi RC(logit) for DOLLAR_2

that the linearity assumption of the partial linear-RC for DOLLAR_2 is valid. Thus, I accept the expected constant change of the partial linear-RC for DOLLAR_2, 0.00210 (from Equation 17.4).

5. Alternatively, the quasi-RC method provides a trusty assumption-free estimate of the partial linear-RC for DOLLAR_2, *partial quasi-RC(linear)*, which is defined as the regression coefficient of the simple ordinary regression of the smooth logit predicted RESPONSE on the smooth DOLLAR_2, columns 8 and 5, respectively. The partial quasi-RC(linear) for DOLLAR_2 is 0.00159 (details not shown).

In sum, the quasi-RC methodology provides alternatives that only the data analyst, who is intimate with the data can decide on: a) accept the partial quasi-RC after asserting the variation among slice-level changes in the partial quasi-RC plot as non-random; b) accept the partial linear-RC (0.00210) after the partial quasi-RC plot validates the linearity assumption; c) accept the trusty partial quasi-RC(linear) estimate (0.00159) after the partial quasi-RC plot validates the linearity assumption linearity. Of course, the default alternative is to outrightly accept the partial linear-RC without testing the linearity assumption. Note that the small difference in magnitude between the trusty and the "true" estimates of the partial linear-RC for DOLLAR_2 is not typical, as the next discussion shows.

I calculate the LRM partial quasi-RC(logit) for LSTORD_M, using six slices to correspond to the distinct values of LSTORD_M, in Table 17.7.

1. The partial quasi-RC plot for LSTORD_M of the relationship between the smooth predicted logit RESPONSE and the smooth LSTORD_M, in Figure 17.3, is clearly nonlinear with expected changes in the logit RESPONSE: −0.0032, −0.1618, −0.1067, −0.0678 and 0.0175.

2. The implication is that the linearity assumption for the LSTORD_M does not hold. There is not a constant expected change in the logit RESPONSE as implied by the prescribed interpretation of the partial linear-RC for LSTORD_M, −0.0798 (in Equation 17.4).

3. The secondary implication is that the structural form of LSTORD_M is not correct. The S-shaped nonlinear pattern suggests that quadratic and/or cubic re-expressions of LSTORD_M be tested for model inclusion.

4. Satisfyingly, the partial quasi-RC(linear) value of −0.0799 (from the simple ordinary regression of the smooth predicted logit RESPONSE on the smooth LSTORD_M) equals the partial linear-RC value of −0.0798. The implications are: a) The partial linear-RC provides the *average* constant change in the the logit RESPONSE across the LSTORD_M range of values between 1 to 66; b) The partial quasi-RC provides more accurate reading of the changes, with respect to the six pre-sliced intervals across the LSTORD_M range.

TABLE 17.7

Calculations for LRM Partial Quasi-RC (logit): LSTORD_M

Slice	min_LSTORD_M	max_LSTORD_M	med_LSTORD_M_r	med_LSTORD_M_r+1	change_LSTORD_M	med_lgt_r	med_lgt_r+1	change_lgt	quasi-RC (logit)
1	1	1	.	1	.	.	-3.2332	.	.
2	1	3	1	2	1	-3.2332	-3.2364	-0.0032	-0.0032
3	3	3	2	3	1	-3.2364	-3.3982	-0.1618	-0.1618
4	3	4	3	4	1	-3.3982	-3.5049	-0.1067	-0.1067
5	4	5	4	5	1	-3.5049	-3.5727	-0.0678	-0.0678
6	5	12	5	6	1	-3.5727	-3.5552	0.0175	0.0175

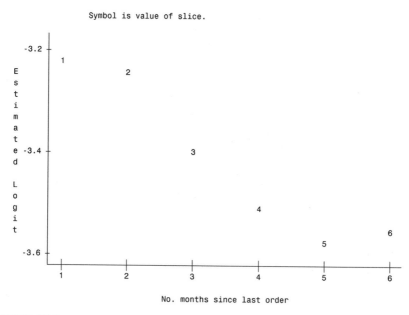

FIGURE 17.3

Visual Display of LRM Partial Quasi RC(logit) for LSTORD_M

Forgoing the details, the LRM partial quasi-RC plots for both RFM_CELL and AGE_Y support the linearity assumption of the partial linear-RC. Thus, the partial quasi-RC(linear) and the partial linear-RC values should be equivalent. In fact, they are: for RFM_CELL, the partial quasi-RC(linear) and the partial-linear RC are −0.2007 and −0.1995, respectively; for AGE_Y, the partial quasi-RC(linear) and the partial-linear RC are 0.5409 and 0.5337, respectively.

In sum, this illustration shows that the workings of the quasi-RC methodology perform quite well on the linear predictions based on multiple predictor variables. Suffice to say, by converting the logits into probabilities — as was done in simple logistic regression illustration in Section 17.3.3 — the quasi-RC approach performs equally well with nonlinear predictions based on multiple predictor variables.

17.4 Quasi-RC for a Coefficient-Free Model

The linear regression paradigm, with nearly two centuries of theoretical development and practical use, has made the equation form — the sum of weighted predictor variables $(Y = b_0 + b_1X_1 + b_2X_2 + ... + b_nX_n)$ — the icon of predictive models. This is why the new machine learning techniques of the last half-century are evaluated by the coefficients they produce. If the new coefficients impart comparable information to the regression coefficient then the new technique passes the first-line of acceptance. If not, the technique is all but summarily rejected. Ironically, some machine learning methods offer better predictions without the use of coefficients. The burden of acceptance of the coefficient-free model lies with the extraction of something familiar and trusting. The quasi-RC procedure provides data analysts and marketers the comfort and security of coefficient-like information for evaluating and using the coefficient-free machine learning models.

Machine learning models without coefficients can assuredly enjoy the quasi-RC method. One of the most popular coefficient-free models is the regression tree, e.g., CHAID. The regression tree has a unique equation form of "If... Then " rules, which has rendered its interpretation virtually self-explanatory, and has freed itself from a burden of acceptance. The need for coefficient-like information was never sought. In contrast, most machine learning methods, like neural networks, have not enjoyed an easy first-line of acceptance. Even their proponents have called neural networks "black boxes." Ironically, neural networks do have "coefficients" (actually, interconnections weights between input and output layers), but no formal effort has been made to translate them into coefficient-like information. The genetic GenIQ Model has no outright coefficients. Numerical values are sometimes part of the genetic model; but they are not coefficient-like in any way, just genetic material evolved as necessary for accurate prediction.

The quasi-RC method as discussed so far works nicely on the linear and nonlinear regression model. In the next section I illustrate how the quasi-RC technique works and how to interpret its results for a quintessential every-model, the nonregression-based, nonlinear and coefficient-free GenIQ Model as presented in Chapter 16. As expected, the quasi-RC technique works with neural network models, and CHAID or CART regression tree models.

17.4.1 Illustration of Quasi-RC for a Coefficient-Free Model

Again, consider the illustration in Chapter 16 of cataloger ABC, who requires a response model to be built on a recent mail campaign. I select the best #3 GenIQ Model (in Figure 16.3) for predicting RESPONSE based on four predictor variables:

1. DOLLAR_2 — Dollars spent within last two years
2. PROD_TYP — Number of different products
3. RFM_CELL — Recency/frequency/money cells (1 = best to 5 = worst)
4. AGE_Y — Knowledge of customer's age (1 = if known; 0 = if not known)

The GenIQ partial quasi-RC(prob) table and plot for DOLLAR_2 are Table 17.8 and Figure 17.4, respectively. The plot of the relationship between the

TABLE 17.8

Calculations for GenIQ Partial Quasi-RC(prob): DOLLAR_2

Slice	min_DOLLAR _2	max_DOLLAR _2	med_DOLLAR _2_r	med_DOLLAR _2_r+1	change_ DOLLAR_2	med_prb_r	med_prb_r+1	change_ prb	quasi-rc (prob)
1	0	50	.	40	.		0.031114713	.	.
2	50	59	40	50	10	0.031114713	0.031117817	0.000003103	0.000000310
3	59	73	50	67	17	0.031117817	0.031142469	0.000024652	0.000001450
4	73	83	67	79	12	0.031142469	0.031154883	0.000012414	0.000001034
5	83	94	79	89	10	0.031154883	0.031187925	0.000033043	0.000003304
6	94	110	89	102	13	0.031187925	0.031219393	0.000031468	0.000002421
7	110	131	102	119	17	0.031219393	0.031286803	0.000067410	0.000003965
8	131	159	119	144	25	0.031286803	0.031383536	0.000096733	0.000003869
9	159	209	144	182	38	0.031383536	0.031605964	0.000222428	0.000005853
10	209	480	182	253	71	0.031605964	0.03208591 6	0.000479952	0.000006760

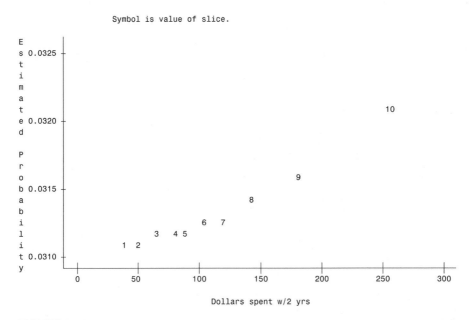

FIGURE 17.4
Visual Display of GenIQ Partial Quasi RC(prob) for DOLLAR_2

smooth predicted probability RESPONSE (GenIQ-converted probability score) and the smooth DOLLAR_2 is clearly nonlinear, which is considered reasonable, due to the inherently nonlinear nature of the GenIQ Model. The implication is that partial quasi-RC(prob) for DOLLAR_2 reliably reflects the expected changes in probability RESPONSE. The interpretation of the partial quasi-RC(prob) for DOLLAR_2 is as follows: for slice 2, which has minimum and maximum DOLLAR_2 values of 50 and 59, respectively, the partial quasi-RC (prob) is 0.000000310. This means that for each and every unit-change in DOLLAR_2 between 50 and 59, the expected constant change in the probability RESPONSE is 0.000000310. Similarly, for slices 3, 4, ..., 10, the expected constant changes in the probability RESPONSE are 0.000001450, 0.000001034, ..., 0.000006760, respectively.[1]

The GenIQ partial quasi-RC(prob) table and plot for PROD_TYP are presented in Table 17.9 and Figure 17.5, respectively. Because PROD_TYP assumes distinct values between 3 and 47, albeit more than a handful, I use twenty slices to take advantage of the granularity of the quasi-RC plotting. The interpretation of the partial quasi-RC(prob) for PROD_TYP can follow the literal rendition of "for each and every unit-change" in PROD_TYP as done for DOLLAR_2. However, as the quasi-RC technique provides alternatives the following interpretations are also available:

---

[1] Note that the maximum values for DOLLAR_2 in Tables 17.6 and 17.8 are not equal. This is because they are based on different M-spread common regions, as the GenIQ Model and LRM use different variables.

TABLE 17.9

Calculations for GenIQ Partial Quasi-RC(prob): PROD_TYP

Slice	min_PROD _TYP	max_PROD _TYP	med_PROD _TYP_r	med_PROD _TYP_r+1	change_ PROD_TYP	med_prb_r	med_prb_ r+1	change_prob	quasi RC (prob)
1	3	6	.	6	.	.	0.031103	.	.
2	6	7	6	7	1	0.031103	0.031108	0.000004696	0.000004696
3	7	8	7	7	0	0.031108	0.031111	0.000003381	.
4	8	8	7	8	1	0.031111	0.031113	0.000001986	0.000001986
5	8	8	8	8	0	0.031113	0.031113	0.000000000	.
6	8	9	8	8	0	0.031113	0.031128	0.000014497	.
7	9	9	8	9	1	0.031128	0.031121	-0.000006585	-0.000006585
8	9	9	9	9	0	0.031121	0.031136	0.000014440	.
9	9	10	9	10	1	0.031136	0.031142	0.000006514	0.000006514
10	10	11	10	10	0	0.031142	0.031150	0.000007227	.
11	11	11	10	11	1	0.031150	0.031165	0.000015078	0.000015078
12	11	12	11	12	1	0.031165	0.031196	0.000031065	0.000031065
13	12	13	12	12	0	0.031196	0.031194	-0.000001614	.
14	13	14	12	13	1	0.031194	0.031221	0.000026683	0.000026683
15	14	15	13	14	1	0.031221	0.031226	0.000005420	0.000005420
16	15	16	14	15	1	0.031226	0.031246	0.000019601	0.000019601
17	16	19	15	17	2	0.031246	0.031305	0.000059454	0.000029727
18	19	22	17	20	3	0.031305	0.031341	0.000036032	0.000012011
19	22	26	20	24	4	0.031341	0.031486	0.000144726	0.000036181
20	26	47	24	30	6	0.031486	0.031749	0.000262804	0.000043801

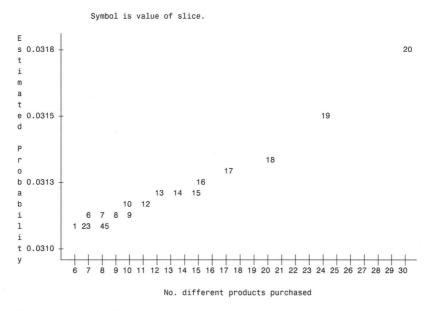

FIGURE 17.5
Visual Display of GenIQ Partial Quasi RC(prob) for PROD_TYP

1. The partial quasi-RC plot of the relationship between the smooth predicted probability RESPONSE and the smooth PROD_TYP suggests two patterns. Pattern one, for PROD_TYP values between 6 and 15, the unit-changes in probability RESPONSE can be viewed as sample variation masking an expected constant change in probability REPSONSE. The "masked" expected constant change can be determined by the average of the unit-changes in probability RESPONSE corresponding to PROD_TYP values between 6 and 15. Pattern two, for PROD_TYP values greater than 15, the expected change in probability RESPONSE is increasing in a nonlinear manner, which follows the literal rendition of "for each and every unit-change" in PROD_TYP.

2. If the data analyst comes to judge the details in partial quasi-RC(prob) table or plot for PROD_TYP as much ado about sample variation, then the partial quasi-RC(linear) estimate can be used. Its value of 0.00002495 is obtained from the regression coefficient from the simple ordinary regression of the smooth predicted RESPONSE on the smooth PROD_TYP (columns 8 and 5, respectively).

The GenIQ partial quasi-RC(prob) table and plot for RFM_CELL are Table 17.10 and Figure 17.6, respectively. The partial quasi-RC of the relationship between the smooth predicted probability RESPONSE and the smooth RFM_CELL suggests an increasing expected change in probability. Recall

TABLE 17.10

Calculations for GenIQ Partial Quasi-RC(prob): RFM_CELL

Slice	min_RFM_CELL	max_RFM_CELL	med_RFM_CELL_r	med_RFM_CELL_r+1	change_RFM_CELL	med_prb_r	med_prb_r+1	change_prb	quasi-RC (prob)
1	1	3	.	2	.	.	0.031773	.	.
2	3	4	2	3	1	0.031773	0.031252	−0.000521290	−0.000521290
3	4	4	3	4	1	0.031252	0.031137	−0.000114949	−0.000114949
4	4	4	4	4	0	0.031137	0.031270	0.000133176	.
5	4	5	4	5	1	0.031270	0.031138	−0.000131994	−0.000131994
6	5	5	5	5	0	0.031138	0.031278	0.000140346	.

TABLE 17.11

Calculations for GenIQ Partial Quasi-RC(prob): AGE_Y

Slice	min_AGE_Y	max_AGE_Y	med_AGE_Y_r	med_AGE_Y_r+1	change_AGE_Y	med_prb_r	med_prb_r+1	change_prb	quasi-RC (prob)
1	0	1	.	1	.	.	0.031177	.	.
2	1	1	1	1	0	0.031177	0.031192	0.000014687	.
3	1	1	1	1	0	0.031192	0.031234	0.000041677	.

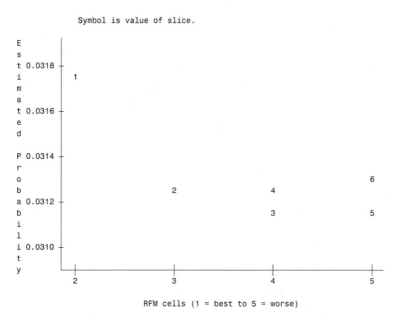

FIGURE 17.6
Visual Display of GenIQ Partial Quasi RC(prob) for RFM_CELL

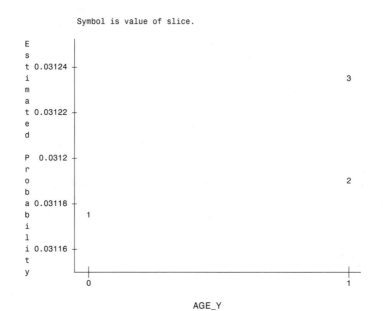

FIGURE 17.7
Visual Display of GenIQ Partial Quasi RC(prob) for AGE_Y

that RFM_CELL is treated as an interval-level variable with a "reverse" scale: 1 = best to 5 = worst; thus, the RFM_CELL clearly has expected nonconstant change in probability. The plot has *double* smooth points at both RFM_CELL = 4 and RFM_CELL = 5, for which the *double-smoothed* predicted probability RESPONSE is taken as the average of the reported probabilities. For RFM_CELL = 4, the twin points are 0.031252 and 0.031137; thus the double-smoothed predicted probability RESPONSE is 0.311945. Similarily, for RFM_CELL = 5, the double-smoothed predicted probability RESPONSE is 0.31204. The interpretation of the partial quasi-RC(prob) for RFM_CELL can follow the literal rendition of "for each and every unit-change" in RFM_CELL.

The GenIQ partial quasi-RC(prob) table and plot for AGE_Y are Table 17.11 and Figure 17.7, respectively. The partial quasi-RC plot of the relationship between the smooth predicted probability RESPONSE and the smooth RFM_CELL is an uninteresting expected linear change in probability. The plot has double smooth points at both AGE_Y = 1, for which the double-smoothed predicted probability RESPONSE is taken as the average of the reported probabilities. For AGE_Y = 1, the twin points are 0.031234 and 0.031192; thus the double-smoothed predicted probability RESPONSE is 0.31213. The interpretation of the partial quasi-RC(prob) for AGE_Y can follow the literal rendition of "for each and every unit-change" in AGE_Y.

In sum, this illustration shows how the quasi-RC methodology works on a nonregression-based, nonlinear and coefficient-free model. The quasi-RC procedure provides data analysts and marketers with the sought after comfort and security of coefficient-like information for evaluating and using coefficient-free machine learning models like GenIQ.

17.5 Summary

The redoubtable regression coefficient enjoys everyday use in database marketing analysis and modeling. Data analysts and marketers use the regression coefficient when interpreting the tried and true regression model. I restated that the reliability of regression coefficient is based on the workings of the linear statistical adjustment. This removes the effects of the other variables from the dependent variable and the predictor variable, producing a linear relationship between the dependent variable and the predictor variable.

In absence of another measure, data analysts and marketers use the regression coefficient to evaluate new modeling methods. This leads to a quandary, as some of the newer methods have no coefficients. As a counterstep, I presented the quasi-regression coefficient (quasi-RC) that provides information similar to the regression coefficient for evaluating and using coefficient-free models. Moreover, the quasi-RC serves as a trusty assumption-free alternative to the regression coefficient, when the linearity assumption is not met.

I provided illustrations with the simple one-predictor variable linear regression models to highlight the importance of the satisfaction of linearity assumption for accurate reading of the regression coefficient itself, as well as its effect on the model's predictions. With these illustrations, I outlined the method for calculating the quasi-RC. Comparison between the actual regression coefficient and the quasi-RC showed perfect agreement, which advances the trustworthiness of the new measure.

Then, I extended the quasi-RC for the everymodel, which is any linear or nonlinear regression, or any coefficient-free model. Formally, the partial quasi-RC for predictor variable X is the expected change — not necessarily constant — in the dependent variable Y associated with a unit-change in X when the other variables are held constant. With a multiple logistic regression illustration, I compared and contrasted the logistic partial linear-RC with the partial quasi-RC. The quasi-RC methodology provided alternatives that only the data analyst, who is intimate with the data can decide on: a) accept the partial quasi-RC if the partial quasi-RC plot produces a perceptible pattern; b) accept the logistic partial linear-RC if the partial quasi-RC plot validates the linearity assumption; and c) accept the trusty partial quasi-RC(linear) estimate if the partial quasi-RC plot validates the linearity assumption. Of course, the default alternative is to accept the logistic partial linear-RC outright without testing the linearity assumption.

Last, I illustrated the quasi-RC methodology for coefficient-free GenIQ Model. The quasi-RC procedure provided me with sought after comfort and security of coefficient-like information for evaluating and using the coefficient-free GenIQ Model.

Index